FGF Signalling in Vertebrate Development

Colloquium Series on Developmental Biology

Developmental biology is in a period of extraordinary discovery and research in this field will have a broad impact on the biomedical sciences in the coming decades. Developmental biology is interdisciplinary and involves the application of techniques and concepts from genetics, molecular biology, biochemistry, cell biology, and embryology to attack and understand complex developmental mechanisms in plants and animals, from fertilization to aging. Many of the same genes that regulate developmental processes underlie human regulatory gene disorders such as cancer and serve as the genetic basis of common human birth defects. An understanding of fundamental mechanisms of development is providing a basis for the design of gene and cellular therapies for the treatment of many human diseases. Of particular interest is the identification and study of stem cell populations, both natural and induced, which is opening new avenues of research in development, disease, and regenerative medicine. This eBook series is dedicated to providing mechanistic and conceptual insight into the broad field of developmental biology. Each issue is intended to be of value to students, scientists, and clinicians in the biomedical sciences.

FGF Signalling in Vertebrate Development
Mary Elizabeth Pownall and Harry V. Isaacs
www.morganclaypool.com

ISBN: 9781615040636 paperback

ISBN: 9781615040643 ebook

DOI: 10.4199/C00011ED1V01Y201004DEB002

A Publication in the Morgan & Claypool Life Sciences Publishers series

DEVELOPMENTAL BIOLOGY

Book #2

Series Editor: Daniel S. Kessler, Ph.D., University of Pennsylvania

Series ISSN Pending

FGF Signalling in Vertebrate Development

Mary Elizabeth Pownall and Harry V. Isaacs
University of York

DEVELOPMENTAL BIOLOGY #2

ABSTRACT

The fibroblast growth factors (FGFs) represent one of the relatively few families of extracellular signalling peptides that have been shown in recent decades to be key regulators of metazoan development. FGFs are required for multiple processes in both protostome and deuterostome groups. Given the wide range of regulatory roles attributed to the FGFs, it is perhaps not surprising that misregulation of this signalling pathway has been implicated in a number of human disease conditions. The focus of the present review is to look at the fundamental components of the FGF pathway and illustrate how this highly conserved regulatory cassette has been deployed to regulate multiple, diverse processes during vertebrate development. This review will explore examples from several vertebrate model organisms and include discussions of the role of FGF signalling in regulating the establishment of the mesoderm, neural patterning, morphogenesis, myogenesis, limb development, and the establishment of right–left asymmetry.

KEYWORDS

fibroblast growth factor, mesoderm, neurectoderm, Xenopus, organogenesis, somites, limb, MyoD, somitogenesis, morphogenesis, left-right asymmetry, MAPK, Sprouty, ERK, Map kinase phosphatase, signal transduction

Contents

Introduction

This review will address the role of fibroblast growth factor (FGF) signalling during vertebrate development, from its early functions in patterning the mesoderm and influencing cell behaviour during morphogenesis to its later roles in limb development and neural patterning in the brain. Beyond the scope of this review, FGFs continue to be important in adult homeostasis, for angiogenesis and for wound healing. There is also considerable interest in FGF regulation of stem cell properties in culture: FGF is a common component in the culture medium of human embryonic stem cells and induced pluripotent stem cells. There is also evidence that FGF signalling enables the differentiation of mouse embryonic stem cells into a number of cell lineages (Kunath et al., 2007; Levenstein et al., 2006; Takahashi et al., 2007). Furthermore, given that the FGF signalling pathway has a fundamental role in cell behaviour and differentiation, it is perhaps not surprising that there is increasing evidence linking disregulation of this pathway to a wide range of pathologies, including skeletal dysplasias, neurodegenerative disease, metabolic disorders and cancer (Krejci et al., 2009; Turner and Grose, 2010).

FGF Ligands

FGF protein was first described when it was purified from bovine brain and found to be mitogenic for cultured fibroblasts (Esch et al., 1985; Gospodarowicz, 1975; Gospodarowicz et al., 1977; Gospodarowicz and Moran, 1975). We now know that FGFs are expressed widely during vertebrate development where they play many diverse and essential roles. FGF ligands are small polypeptides with a partially conserved core of 120–130 amino acids that bind heparin with high affinity. By this criteria, there are 22 FGF ligands in mammals; however, FGF11–14 do not activate FGF receptors, so perhaps these should not be considered FGF ligands (Beenken and Mohammadi, 2009). Phylogenetic analysis of human FGF ligands reveals groups of more related members and divides them into seven subfamilies (Itoh and Ornitz, 2004) (Table 1). Five of these FGF subfamilies are characterised by paracrine signalling molecules that signal by forming a tripartite complex with FGF receptors and heparan sulphate glycosaminoglycan (HS-GAG) chains. The other two groups, FGF11–14, and FGF19, 21, and 23, are different. FGF11–14 act intracellularly, while FGF19, 21 and 23 have reduced HS binding affinity and function in an endocrine manner to impact on adult homeostasis and metabolism (Beenken and Mohammadi, 2009; Ornitz, 2000). Noncanonical FGF ligands have also been reported to activate signalling through the FGF receptors; these include a number of cell adhesion molecules, such as neural cell adhesion molecule (Hall et al., 1996).

TABLE 1:

FGF SUBFAMILY	LIGANDS	RECEPTOR PREFERENCE
FGF1	FGF1, FGF2	FGF1 activates all FGFRs; FGF2 prefers FGFR1c and FGFR2c
FGF4	FGF4, FGF5, FGF6	FGFR1c, FGFR2c
FGF7	FGF3, FGF7, FGF10, FGF22	FGFR2b, FGFR1b
FGF8	FGF8, FGF17, FGF18	FGFR3c, FGFR4, FGFR1c
FGF9	FGF9, FGF16, FGF20	FGFR3c, FGFR2c
FGF19	FGF19, FGF21, FGF23	Hormone class, very weak activation of FGFR1c, FGFR2c
FGF11	FGF11, FGF12, FGF13, FGF14	No activation of FGFRs

Phylogenetic alignment of human FGF ligands has classified the FGF ligands into seven subfamilies. The members of five of these FGF subfamilies are paracrine signaling molecules that signal by forming a tripartite complex with FGF receptors and heparan sulphate glycosaminoglycan (HS-GAG) chains. FGF11-14 do not bind or activate FGFRs, while FGF19, FGF21, and FGF23 function in an endocrine manner (see Itoh and Ornitz, 2004 and Zhang et al., 2006).

FGF Receptors

Four genes code for FGF receptors (FGFRs), which are receptor tyrosine kinases (RTKs) and are activated by ligand binding. FGFRs have three immunoglobulin-like domains Ig1, Ig2, and Ig3 that bind ligand, HS-GAG, and are important in receptor dimerisation. Alternative splicing of part of the Ig3 domain generates different isoforms of FGFR1–3 with distinct binding specificities (Johnson et al., 1991) that are expressed in different tissues (Orr-Urtreger et al., 1993). Additionally, an individual FGF ligand may have a preference for a particular receptor. Using a mitogenic assay in cultured cells expressing distinct FGFRs, only FGF1 was shown to promiscuously activate all four FGFRs while the other ligands were more effective on cells expressing different receptors. For instance, the FGF4 subfamily is more active on cells expressing FGFR1c or 2c, while the FGF7 (see Table 1) subfamily is more active on cells expressing FGFR1b or 2b (Zhang et al., 2006). Another splice variant in FGFR1 has two amino-acid deletion of Val(423)–Thr(424) in the juxtamembrane region and shows different expression and activity than the isoform that includes these two amino acids (Paterno et al., 2000).

An in vivo study has shown that overexpressing constitutively active or dominant negative forms of different FGFRs in zebrafish embryos will generate different phenotypes (Ota et al., 2009). Overall, it was found that all of the constitutively active FGFRs tested gave rise to posteriorised/dorsalised embryos that lack forebrain structures and all of the dominant negative forms cause posterior truncations in zebrafish as described previously for *Xenopus* (Amaya et al., 1991). However, the potency of different receptors and their ability to affect the activity of individual FGF ligands varied. The dominant negative FGFRs (dnFGFR) were constructed by deleting the cytoplasmic kinase domains. When the dnFGFR are overexpressed in the embryo, the mutant receptors partner with endogenous wild-type FGFRs, but because the activation of signal transduction relies on cross-phosphorylation of the intracellular kinase domains, these mutant/wt FGFR dimers will be nonfunctional. A dominant negative form of FGFR1 (dnFGFR1) has been widely used in developmental biology (Amaya et al., 1991; Amaya et al., 1993; Branney et al., 2009; Griffin et al., 1995; Isaacs et al., 1994; Pownall et al., 1998) and has been shown to completely block the activation of ERK in early frog embryos (Christen and Slack, 1999). Although overexpressing a dominant negative form of FGFR4 (dnFGFR4) was found to give rise to a different phenotype than dnFGFR1 (Hardcastle and Papalopulu, 2000; Hongo et al., 1999), it is known that these receptors will heterodimerise promiscuously (Ueno et al., 1992) and therefore the specificity of the assay is lost. This has been shown recently for dnFGFR1 and dnFGFR4 in a microarray study (Branney et al., 2009).

In addition to the four RTKs, another member of the FGFR family has been identified in a number of species (Sleeman et al., 2001). The extracellular domain of FGFRL1 shares about 50% amino acid identity with the other FGFRs. As with FGFR1–4, FGFRL1 has been shown to have differing affinity for the various FGF ligands (Bertrand et al., 2009; Trueb et al., 2005); however, FGFRL1 lacks any intracellular tyrosine kinase activity. It has recently been shown that FGFRL1 is shed from cells and when overexpressed in *Xenopus* embryos can mimic the morphological effects of FGF inhibition (Steinberg et al., 2010).

Heparan Sulphate Proteoglycans

FGF proteins are characterised by their ability to bind heparin, which is a highly sulphated form of heparan sulphate produced only by mast cells and well known for its activity as an anticoagulant. The high affinity of FGFs for heparin was key for the purification of FGF proteins. Heparan sulphate proteoglycans (HSPGs) are produced by virtually all cells and are present on the cell surface and in the extracellular matrix. They consist of a protein core to which long, unbranched chains of repeating disaccharides are attached. Heparan sulphate GAG chains are made up of the sugar subunits glucoronic acid and *N*-acetyl glucosamine; these subunits are highly modified by sulfation and epimerisation by specific enzymes present in the golgi. *N*-Deacetylase/*N*-sulfotransferase replaces the *N*-acetyl group of *N*-acetyl glucosamine with a sulphate group and then the glucuronic acid residues adjacent to N-sulphated glucosamine are epimerised to iduronic acid. After this, 2-*O*-sulfotransferase acts on iduronic acid and 6-*O*-sulfotransferase (6OST) on the glucosamine; a few are also modified by 3-*O*-sulfotransferase. However, not all of the available substrate is sulphated, which generates regions of high and low sulfation along the HS chain and results in an enormous amount of structural diversity in HSPGs. This rich heterogeneity allows HSPGs to bind many different proteins and among these proteins are growth factors and receptors including FGFs and FGFRs.

HSPGs are essential for FGF signalling in vivo (Lin et al., 1999). The crystal structure of FGF ligand bound to FGFR reveals heparin bridging to contact receptors and ligands, cross-linking and stabilising the signalling complex (Schlessinger et al., 2000). The 6-*O*-sulphate group on the glucosamine in heparin was found to be central in these interactions (Pye and Gallagher, 1999; Turnbull et al., 2001). This is important because enzymes that modify 6-O-sulfation of HSPGs have been found to play a role in the modulation of FGF signalling (Freeman et al., 2008; Sugaya et al., 2008; Wang et al., 2004).

FGF Signal Transduction

Assembly of the FGF signalling complex following FGF and HS-GAG binding results in receptor dimerisation and the activation of intracellular signal transduction pathways (reviewed (Beenken and Mohammadi, 2009; Bottcher and Niehrs, 2005; Knights and Cook, 2010; Thisse and Thisse, 2005). Some of the pathways activated downstream of FGF are illustrated in Figure 1. Activation of FGF signalling leads to phosphorylation of a number of conserved tyrosine residues in the intracellular domain of the FGFR. An important target of FGFR phosphorylation is the FGFR substrate 2 (FRS2) (Kouhara et al., 1997) which associates with the receptor and in turn allows the recruitment of the Grb2 adaptor protein and the associated nucleotide exchange factor son of sevenless (SOS) (Ong et al., 2000). Grb2/SOS then activates the small GTP binding protein Ras by stimulating it to cycle from its inactive GDP bound to its active GTP bound form. Activation of Ras leads to the stimulation of a cascade of phosphorylation events involving Raf (a MAPK kinase kinase) and Mek (MAPK kinase, ultimately leading the activation and phosphorylation of MAPK ERK. Phosphorylated MAPK, a serine/threonine kinase, is then able to phosphorylate and modify the activity transcription factors. Notable among the transcription factors activated by MAPK are ETS proteins (Randi et al., 2009) that have a winged helix-turn-helix protein fold as a DNA binding domain and have been shown to be important effectors of FGF signalling, regulating gene expression downstream of the MAPK pathway (Nentwich et al., 2009). The genes transcribed in response to ETS proteins and other transcription factors activated by this pathway are considered FGF target genes.

Acting via the GAB1 scaffolding protein associated with Grb2, FGF signalling can also activate the phosphoinositide-3 kinase pathway that leads to the generation various phosphoinositdes, which are phosphorylated on their 3′ carbon atom. An important downstream mediator of this pathway is the serine/threonine kinase Akt/protein kinase B, which has multiple roles in regulating cell survival and growth in normal development and in the context of cancer (Nicholson and Anderson, 2002).

Activation of phospholipase Cγ (PLCγ) by FGF signalling is dependent on recruitment of the enzyme to a conserved phosphotyrosine residue in the FGFR (Mohammadi et al., 1991; Ueda et al., 1996). PLC hydrolyzes phosphotidylinositol-4,5-diphosphate to inositol-1,4,5-trisphosphate (IP3) and diacylglycerol (DAG). DAG activates protein kinase C, while IP3 stimulates intracellular calcium release.

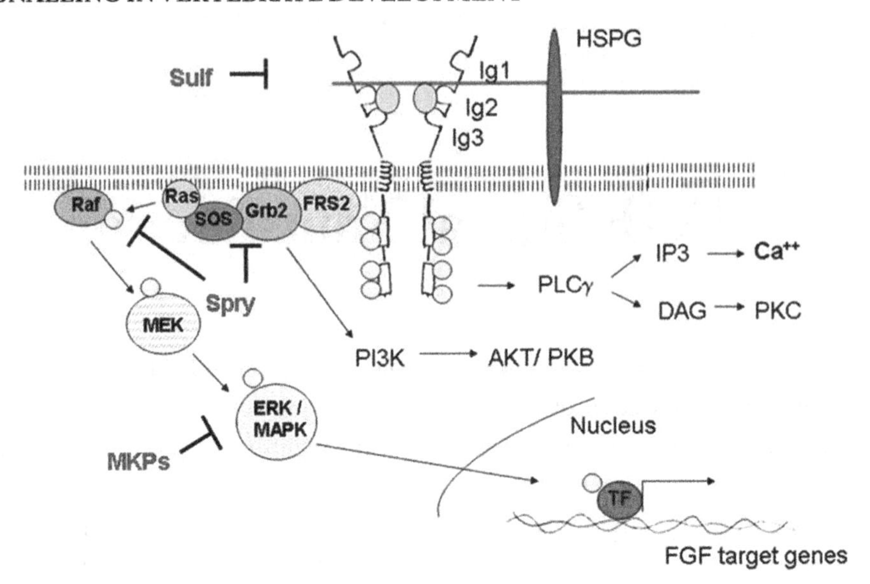

FIGURE 1: FGF signal transduction. Formation of the FGF:FGFR:HS signalling complex causes the activation of the intracellular kinase domains and the cross-phosphorylation of tyrosines on the FGFRs. FRS2 interacts with the phosphorylated tyrosines and is phosphorylated itself. FRS2 then activates the adaptor protein Grb2 that associates with SOS, a nucleotide exchange factor which activates Ras. Ras is a small GTP binding protein that activates Raf, which activates MEK which activates MAPK (ERK). FRS2 activity also activates phosphoinositide-3 kinase, which activates AKT/PKB. PLCγ binds the actvated FGFR by its SH2 domain and then generates inositol-1,4,5-trisphosphate and DAG from phosphotidylinositol-4,5-diphosphate resulting in the activation of protein kinase C and the release of intracellular calcium. The FGF pathway is negatively regulated by Sulf, Spry, and MAPK phosphatases.

Regulators

SULFATASES AND SULFOTRANSFERASES

As discussed above, HS-GAG chains are a central part of the FGF signalling complex (Figure 2). Moreover, the sulphate group at the 6-O position of glucosamine has been shown to be essential for the association of ligand and receptor, making contact with FGFR (Pellegrini, 2001). The contact of the HS-GAGs with FGFR and FGF ligand draws the signalling complex together and stabilizes it (Schlessinger et al., 2000). In addition, the presence of the 6-O sulphate group also has been shown to be critical for the mitogenic activity of FGF (Lundin et al., 2000). It is therefore not surprising that enzymes controlling the presence of the 6-O sulphate group can influence FGF signalling.

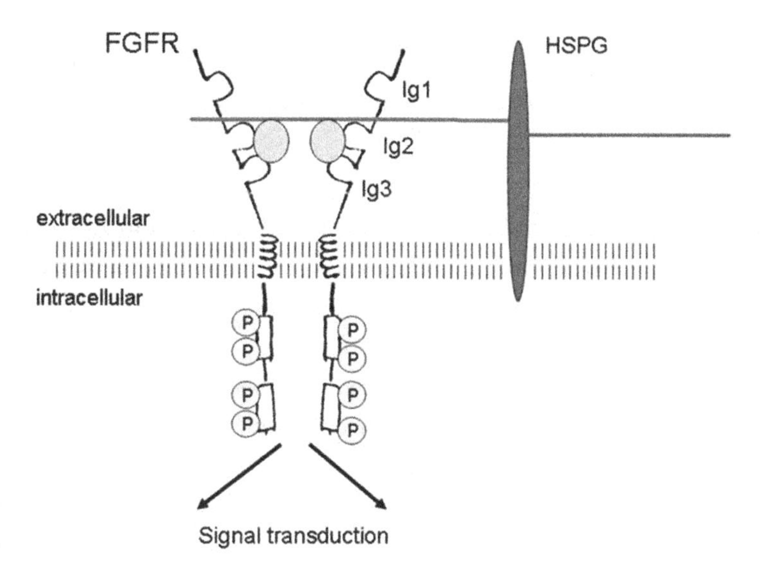

FIGURE 2: The FGF signalling complex. FGF ligands (shown as light blue ovals) interact with extracellular immunoglobulin domains of the FGF receptors. The GAG chains of HSPGs make contact with both the FGF ligands and the FGF receptors driving the formation of the tripartite FGF signalling complex. The dimerisation of FGF receptors results in the auto- and cross-phosphorylation of tyrosines in the intracellular kinase domains and sets off signal transduction cascades.

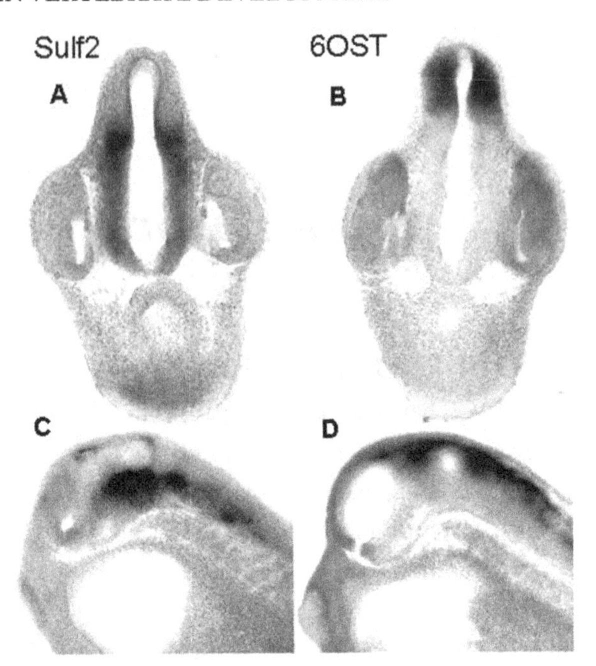

FIGURE 3: Sulf2 and 6OST are expressed in complementary regions of the developing *Xenopus* brain. Vibratome sections subsequent to in situ hybridisation (ISH) shows that the expression of Sulf2 is restricted to the ventral part of the midbrain and hindbrain, while the expression of 6OST is restricted dorsally. A and B are cross sections; C and D are sagittal sections through tailbud stage 28 embryos. Adapted from Winterbottom and Pownall, 2008.

During the biosynthesis of HSPGs, 6OST acts in the Golgi to add 6-*O*-sulphate groups to glucosamine in some regions of the HS chain; however, not all disaccharides are modified, leading to structural diversity along the HS chain (Turnbull et al., 2001). Further modification can come subsequent to HSPG synthesis: the enzymes Sulf1 and Sulf2 are HS-6-*O*-endosulfatases that remove the 6-*O*-sulphate group from HS chains (Dhoot et al., 2001; Morimoto-Tomita et al., 2002). There are some excellent examples of the complementary expression of these enzymes with opposite activities (Figure 3), suggesting that the 6-O modification of HSPGs is very important for developmental patterning. In vivo studies have found that *Drosophila* lacking *6OST* have defective trachea (Kamimura et al., 2001; Kamimura et al., 2006) which is known to require FGF during its morphogenesis. *6OST* genes have distinct expression patterns during zebrafish development (Cadwallader and Yost, 2006), and zebrafish *6OST* mutants have abnormalities in the development of skeletal muscle and blood (Bink et al., 2003; Chen et al., 2005).

It is well established that Sulf1 and Sulf2 are negative regulators of FGF signalling (Freeman et al., 2008; Lai et al., 2003; Wang et al., 2004). Overexpression of *Sulf1* inhibits the activation of dpERK in *Xenopus* explants treated with FGF and in ovarian cancer cells. Knockdown of *Sulf1* up-regulates dpERK in vivo during *Xenopus* development, indicating Sulf1 is an endogenous regulator of FGF signalling. Gene targeting of *Sulf1* and *Sulf2* increases the response of mouse embryonic fibroblasts (MEFs) to FGF (Lamanna et al., 2008). Interestingly, the expression of the 6OST genes is up-regulated in the *Sulf1/2* knockout MEFs, suggesting a feedback mechanism where Sulf activity represses *6OST* gene expression (Lamanna et al., 2008). MEFs where *6OST* has been knocked out have a reduced response to FGF (Sugaya et al., 2008), but any effect on *Sulf* gene expression has not been determined.

SPROUTY

Sprouty (Spry) was first identified as a *Drosophila* mutant (Hacohen et al., 1998) that had high levels of FGF activity, indicating that the normal function of the *Sprouty* gene is to restrict FGF signalling, and RTK signalling in general (Casci et al., 1999). In mammals, there are four Spry homologues and all possess a cysteine-rich Spry domain in the C-terminal half of the protein. At the N-terminus, there is a conserved tyrosine residue that is essential for the inhibitory function of the Spry proteins (see reviews of Cabrita and Christofori, 2008; Mason et al., 2006). Mutating this tyrosine results in a dominant negative Spry that enhances MAPK signalling downstream of FGF (Hanafusa et al., 2002), and it has been suggested that this tyrosine is needed for Spry to complex with other proteins to efficiently inhibit MAPK signalling (Mason et al., 2006). Sprouty proteins can act at multiple levels to restrict the MAPK signalling pathway. Spry has been shown to act upstream of Ras and compete with FRS2 for binding to Grb2/SOS complex (Hanafusa et al., 2002), while other studies have shown that Spry proteins inhibit Raf activation (Sasaki et al., 2003). Importantly, the expression of *Spry* genes is activated by FGF signalling (Branney et al., 2009; Hacohen et al., 1998; Sivak et al., 2005), and this creates a negative feedback loop whereby FGF restricts its own activity by activating the expression of *Spry*. Spry proteins are also regulated posttranscriptionally by FGF where phosphorylation of the conserved N-terminal tyrosine is required for its ability to inhibit the MAPK pathway (Mason et al., 2004). In addition, Spry proteins are phosphorylated at two serine residues that are dephosphorylated in response to FGF, activating Spry. These same serine residues are recognised by the ubiquitin ligase CBL, which competes with the phosphatase activated by FGF signalling, such that Spry is either dephosphorylated and activated by FGF or targeted to the proteasome by CBL (Lao et al., 2007).

MAPK PHOSPHATASES

Activation of the MAPK pathway by FGF results in a series of phosphorylation events that are reversible. The reversible nature of ERK phosphorylation means that its activity relies on a balance between the kinase that phophorylates it (MEK) and the phophatase that dephosphorylates it.

MAPK phosphatases (MKPs) are dual specificity phosphatases that target and dephosphorylate activated MAPKs. There are several MAPKs including ERK, JNK, and p38 that are inactivated by specific MKPs/dual-specificity phosphatases (DUSPs). dpERK is inactivated when the phosphorylation of the tyrosine or the threonine within its activation loops is removed by a DUSP. There are 11 genes coding for DUSPs in mammalian genomes (Alonso et al., 2004) and some are highly specific to their MAPK substrate; for instance, MKP3/DUSP6 (also known as Pyst1) specifically inactivates dpERK (Keyse, 2000). The binding of MKP3 to dpERK stimulates the phosphatase activity of MKP; the activation and nuclear translocation of ERK is blocked. In this way, MKPs/DUSPs function as intracellular brakes of FGF signal transduction (Eblaghie et al., 2003).

Interesting, the expression of some MKPs/DUSPs were found in regions of developing embryos known to be active in FGF signalling (Dickinson et al., 2002; Eblaghie et al., 2003; Gomez et al., 2005; Lewis et al., 1995; Li et al., 2007). In addition, MKP1, MKP3, and DUSP5 have all been found to require FGF signalling for their expression. Consistent with these findings, the transcriptional activation of the MKP/DUSP genes is an early and robust response to FGF signalling (Branney et al., 2009; Eblaghie et al., 2003). Gene targeting of MKP3/DUSP6 results in increased dpERK levels and dwarfing phenotypes similar to those seen in embryos where strong activating FGFRs are expressed (Li et al., 2007). Overexpression of MKP3 or DUSP5 blocks mesoderm induction by FGF in *Xenopus* tissues (Branney et al., 2009; Umbhauer et al., 1995).

These observations provide another example where FGF signalling activates the expression of a negative regulator of FGF. It is possible that this negative feedback provides a mechanism to control the duration of signal transduction or even create an oscillatory response downstream of FGF.

SEF

Sef (**S**imilar **E**xpression to **F**GF) is a transmembrane protein originally identified in zebrafish and negatively regulates FGF signalling in development and cell culture (Furthauer et al., 2002; Tsang et al., 2002). The Sef protein associates with FGFRs, and it has been reported that the extracellular, transmembrane, and intracellular domains of Sef are involved in FGF antagonism (Ren et al., 2007). However, the exact mode of action of Sef has yet to be elucidated.

FLRT

The three FLRT (**F**ibronection-**L**eucine-**R**ich-**T**ransmembrane) genes were originally identified in human adult skeletal muscle (Lacy et al., 1999). It was subsequently shown *Xenopus* FLRT3 associates with FGFRs and, unlike the previously describe modulators, can act as a positive regulator of FGF signalling and is required for normal responses to FGF signalling in embryo tissues (Bottcher et al., 2004). In contrast, a recent report indicates that transcriptional responses to FGF signalling are normal in FLRT3 knockout mice (Maretto et al., 2008).

The sprouties, MKPs, Sef, and FLRTs are members of the FGF synexpression group that share similar expression patterns with the FGFs and are transcriptionally regulated by FGF signalling (Bottcher et al., 2004). Thus, presence of multiple negative and positive feedback loops operating downstream of FGF signalling illustrates the importance of fine-tuning the levels and extent of FGF signalling in the vertebrate embryo.

Integration with Other Signalling Pathways

FGF signalling is one among a handful of cell signalling pathways that orchestrate embryonic development. Integration of these pathways is central to the correct patterning, cell specification, and tissue differentiation that occur during normal development; many examples of the interplay of FGF and other signalling pathways are presented in this review.

The canonical Wnt signalling pathway, for instance, has several points where cross-talk can take place with the FGF pathway (see Figure 4). Stabilisation of β-catenin is the major result of the canonical Wnt pathway (recent reviews of Wnt signalling (Kikuchi et al., 2009; van Amerongen and Nusse, 2009). In the absence of Wnt signals, β-catenin is phosphorylated by a destruction complex including Axin, APC, and GSK3, which primes it for degradation by the ubiquitin pathway. When Wnt signalling is activated, GSK3 is inhibited by Dishevelled (Dvl) and β-catenin is stabilised, accumulates, and moves to the nucleus to promote transcription by associating with the transcription factor TCF/LEF. In the absence of Wnt signalling, TCF is often associated with the co-repressor Groucho so that TCF-dependent transcription is inhibited. One point where FGF signalling feeds into this pathway is by phosphorylating GSK3 (Dailey et al., 2005). Neural progenitor cells derived from mice were treated with FGF protein and found to have higher levels of phospho-GSK3 and β-catenin (Israsena et al., 2004). This is most likely mediated by FGF activation of AKT (Hashimoto et al., 2002; Jope and Johnson, 2004; Katoh, 2006). However, in other contexts, AKT is not involved in mediating interactions between the FGF and Wnt signalling pathways (Keenan et al., 2006).

Another level of integration occurs in the nucleus: MAPK phosphorylation sites have been identified in the co-repressor Groucho and FGF signalling has been shown to inhibit Groucho's repressor activity in vivo (Cinnamon et al., 2008; Murai et al., 2007). In addition, the ability of FGF to phosphorylate Groucho has been shown to affect transcription downstream of Wnt in cells expressing a TCF-dependent reporter (Burks et al., 2009). One model is that MAPK phosphorylation of Groucho allows it to dissociate with TCF, thus promoting the β-catenin/TCF partnership that drives Wnt dependent transcription. The effects of MAPK phosphorylation of Groucho could

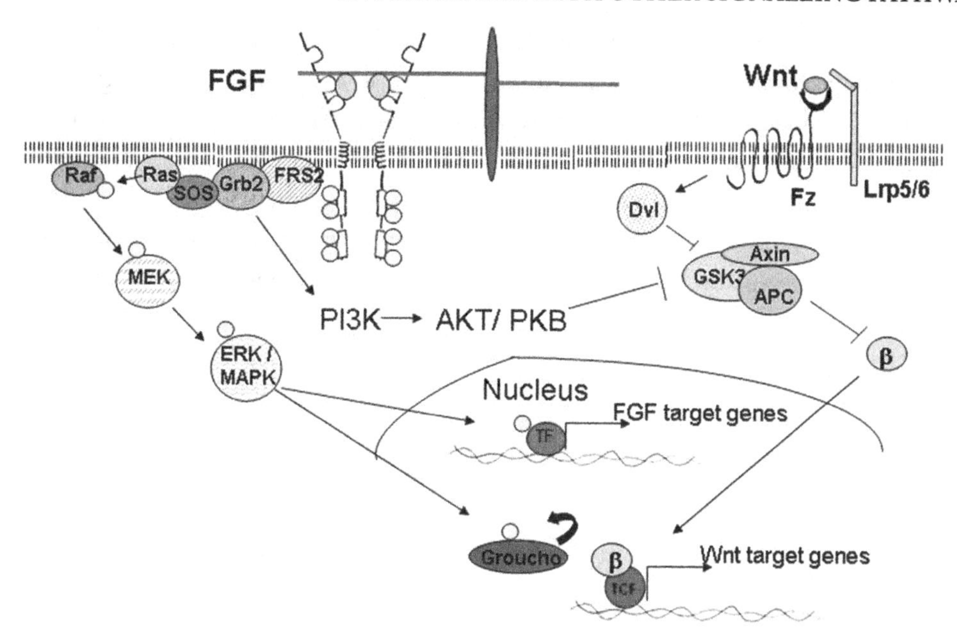

FIGURE 4: Interaction of the FGF and canonical Wnt signalling pathways. Two examples of FGF cross talk with Wnt signalling are shown: AKT, which is activated by FGF, inhibits GSK3 which further represses the β-catenin destruction complex and enhances the stabilisation of β-catenin. MAPK phosphorylation of the transcriptional repressor groucho may release it from its association with TCF, promoting association of TCF with β-catenin to stimulate transcription of Wnt-dependent genes.

facilitate the integration of FGF signalling with many signalling pathways as Groucho is known to operate downstream of Notch, BMPs, TGFb, and Wnts (Hasson and Paroush, 2006).

Integration of Wnt and FGF signalling can also occur at the level of the individual gene promoter. The *Xenopus Cdx4* gene contains functional binding sites for the transcriptional effectors of both the FGF and Wnt signalling pathways (Haremaki et al., 2003).

β-catenin also functions at the cell membrane by associating with cell adhesion proteins including E-cadherin (Willert and Nusse, 1998). Thus, the presence of large amounts of E-cadherin could influence levels of β-catenin available to mediate Wnt signalling by sequestering it at the cell membrane. In FGFR1 mutant mice, high levels of E-cadherin are expressed and β-catenin does not accumulate in the cytoplasm (Ciruna and Rossant, 2001) suggesting another way where FGF can promote Wnt signalling by inhibiting the expression of E-cadherin.

FGF directly interacts with the BMP signal transduction pathway by phosphorylating and inhibiting the activity of SMAD1 (Pera et al., 2003). FGF also impacts BMP signalling through transcriptional regulation: by repressing the expression of BMP ligands (Furthauer et al., 2004) and activating the expression of secreted BMP antagonists (Branney et al., 2009; Fletcher and Harland, 2008). A fully active FGF signalling pathway is essential for activin (a TGFβ signalling molecule, often used to mimic nodal signalling in experiments using *Xenopus* explants) to induce mesoderm (Cornell and Kimelman, 1994), and this dependence could involve the rapid, activin-dependent expression of FGF ligands, such as FGF4 (Fisher, 2002). The interplay of FGF signalling with the other signalling pathways important for embryonic development, including the Wnt, BMP, nodal, and sonic hedgehog pathways, will be evident in the diverse examples discussed in this review.

Mesoderm Induction

Embryonic induction was first described by Hans Spemann in the early 20th century. In his most famous experiment, Spemann grafted the dorsal blastopore lip from a gastrula-staged amphibian embryo onto the ventral side of a differently pigmented host and found that the small graft itself gave rise only to a bit of notochord; however, it induced the surrounding host tissue to form a second axis with a well-patterned neural tube and axial mesoderm. The central role for cell–cell signalling during embryonic development was thus established. Further studies describing the capacity of the vegetal hemisphere of urodele embryos to induce mesoderm in the adjacent marginal zone were carried out by Peter Nieuwkoop in the 1960s and 1970s. He showed that animal pole explants taken from a blastula-stage embryo ("animal caps") cultured on their own give rise to surface ectoderm, however, when combined with vegetal cells will form mesoderm. The mesoderm is derived solely from the animal cap cells and not the vegetal cells, indicating that the vegetal cells were providing a signal to change the fate of the responding animal cap cells to mesoderm. A model well supported by embryological experiments is the three-signal model presented by Jonathan Slack (Slack et al., 1987) and is shown in Figure 5. In this model, there are two signals from the vegetal pole, a ventral signal (VV) and a dorsal signal (DV), which respectively induces extreme ventral and extreme dorsal mesoderm in the overlying marginal zone during blastula stages. A third, dorsalising signal (O) is derived from the dorsal marginal zone (Spemann's organiser), which imparts the full range of mesodermal fates along the dorsoventral axis of the mesoderm during gastrula stages. Although the demonstration that this third interaction is mediated by the interaction of opposing gradients of BMP ligands and BMP antagonists perhaps suggests that early induction and patterning of the mesoderm is best described by a four-signal model.

Two important papers in 1987 demonstrated for the first time that a purified growth factor could mimic the mesoderm-inducing activity of the vegetal pole. Slack et al. (1987) purified bFGF from bovine brain and used it to induce mesoderm in animal caps. They found that low concentrations of bFGF (FGF2) induced mesoderm that had the same character as mesoderm induced by the VV signal. This work also showed that the endogenous mesoderm-inducing signal emanating from vegetal cells could be blocked by heparin, supporting the notion that FGF, a known heparin binding protein, could possibly be one of the endogenous mesoderm-inducing signals. Kimelman and Kirschner (1987) provided further evidence for this by cloning bFGF from a *Xenopus* oocyte library and showing that it is expressed during amphibian development. Much more recent analyses have shown that in the *Xenopus tropicalis* genome, there are genes coding for 20 FGF ligands and at least 13 of these are expressed during embryonic development (Lea et al., 2009).

Clear evidence that FGF signalling plays a role in mesoderm formation was presented in another landmark paper from the Kirschner lab (Amaya et al., 1991) where an FGFR lacking its intracellular kinase domain was overexpressed in *Xenopus* embryos. Because FGF signalling requires receptor dimerisation and subsequent cross-phosphorylation of tyrosine residues, mutant receptors were able to partner with the normal endogenous receptors, providing a true dominant negative method to block all FGF signalling. This resulted in the loss of much of the mesoderm (Amaya et al., 1991). Further work using this powerful tool demonstrated that there is a subset of mesodermal genes that absolutely require FGF signalling for their expression during gastrula stages (Amaya et al., 1993; Isaacs et al., 1994; Schulte-Merker and Smith, 1995), and notable among these genes is the pan-mesodermal marker *Xenopus brachyury* (*Xbra*) (Smith et al., 1991). An essential role for FGF signalling in mesoderm formation and maintaining mesodermal gene expression has also been described in mammals, birds, and fish (Ciruna and Rossant, 2001; Griffin et al., 1995; Mathieu et al., 2004).

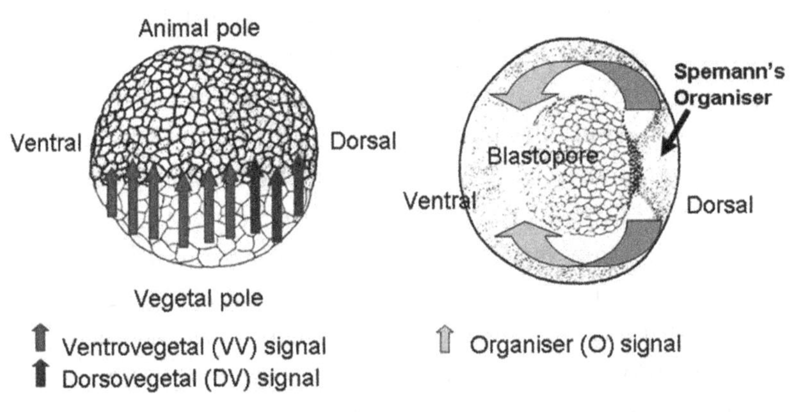

FIGURE 5: The three-signal model for mesoderm induction. In this model, there are two signals from the vegetal pole, a ventral signal (VV) and a dorsal signal (DV), which respectively induce extreme ventral and extreme dorsal mesoderm in the overlying marginal zone during blastula stages. On the left, a sketch of a blastula-stage *Xenopus* embryo is overlaid with arrows depicting these signals. A third, dorsalising signal (O) occurs during gastrula stages and a sketch of the vegetal view of a gastrula embryo is shown. The mesoderm forms a ring around the blastopore. The organiser signal is secreted from the dorsal marginal zone (Spemann's organiser) and its activity allows the development of the full spectrum of mesodermal cell fates along the dorsoventral axis of the mesoderm.

Although these studies indicate an essential role for FGF in mesoderm formation, there is no good evidence that FGF plays a role as an endogenous, vegetally localised mesoderm-inducing factor as had been envisaged by the early papers described above (Kimelman and Kirschner, 1987; Slack et al., 1987). The data on the whole support a model where FGF signalling is required in the marginal zone for it to respond to the vegetally localised mesoderm-inducing factors (see Isaacs, 1997). The endogenous mesoderm-inducing factors are widely accepted to be members of the TFGβ family of signalling molecules, *Xenopus* nodal related (Xnr1, 2, and 4) (Agius et al., 2000; Kofron et al., 1999) and Vg1 (Birsoy et al., 2006; Weeks and Melton, 1987). The signal transduction pathway downstream of Xnrs and Vg1 is often activated in experiments using the commercially available ligand activin. Using this approach, mesoderm induction by activin was found to require a functional FGF signalling pathway. When FGF signal transduction is inhibited using the dominant negative FGFR (or mutant forms of Ras or Raf), mesoderm induction by activin is blocked and a subset of mesodermal genes is not expressed (Cornell and Kimelman, 1994; LaBonne and Whitman, 1994).

These results indicate that FGF is required in the response to mesoderm induction and, indeed, the earliest activation of dpERK can be visualised in blastula-stage embryos in the dorsal marginal zone and animal hemisphere (Branney et al, 2009); this is the tissue responding to mesoderm induction at this time. Later, the activation of dpERK can be visualised in the ring of mesoderm around the closing blastopore in *Xenopus* and in the germ ring during epiboly in zebrafish (Figure 5). While other factors signal through RTKs to activate dpERK, this early MAPK activity has been shown to be completely dependent on FGF signalling (Christen and Slack, 1999), so Figure 6 gives a good picture of FGF activity in the mesoderm. The expression of the ligands *Fgf3*, *Fgf4*, *Fgf8*, and *Fgf20* is found in the early mesoderm (Branney et al., 2009; Christen and Slack, 1997; Isaacs et al., 1995; Lea et al., 2009; Lombardo et al., 1998) and *Fgf4* has been shown to be expressed

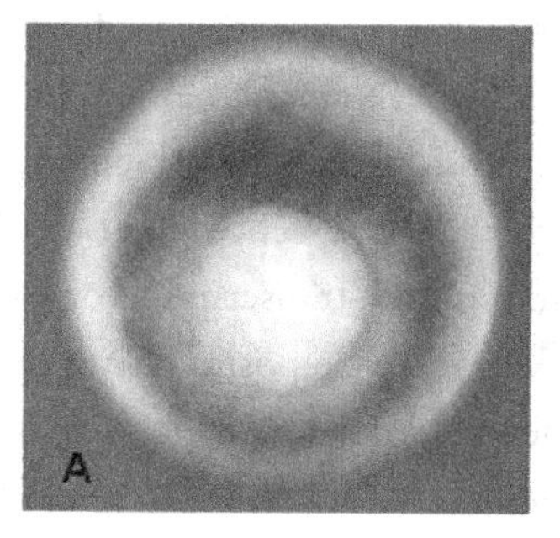

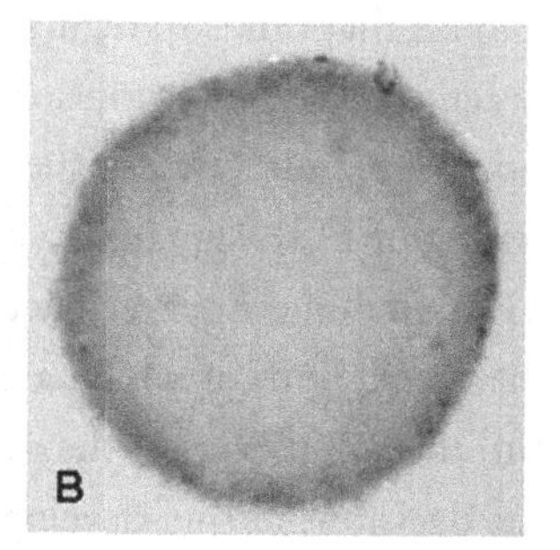

FIGURE 6: dpERK is present in the early mesoderm during gastrulation in *Xenopus* (A) and zebrafish (B). An antibody that specifically recognises the active, diphosphorylated form of MAPK/ERK was used in a whole-mount immunohistochemistry to visualise its localisation during development.

as a direct response to activin, even in the presence of protein synthesis inhibitors (Fisher, 2002). These data point to a model for mesoderm induction where Nodal signalling from the vegetal pole induces mesoderm; and among the earliest genes activated in the nascent mesoderm are *Fgf4* and *Fgf8*, which are required to activate and maintain the expression of a subset of genes important in specifying mesoderm identity (Fletcher and Harland, 2008).

Brachyury is a T-box transcription factor that is transcribed in response to FGF signalling in the absence of protein synthesis indicating that it is a direct and early response. In addition, *Fgf4* can be transcriptionally activated by Xbra itself, while *Xbra* expression requires FGF. This points to an autocatalytic regulatory loop in the early mesoderm where FGF signalling induces the expression of *Xbra*, which in turn feeds back to maintain the expression of *Fgf4* (Isaacs et al., 1994; Schulte-Merker and Smith, 1995). More recently, the initial activation of *Xbra* has been shown to require FGF signalling: the earliest expression of *Xbra* is lost in embryos treated with SU5402. However, the expression of *FGF4* and *FGF8* does not fully depend on the presence of Xbra and are most likely activated by Nodal signalling (Fletcher and Harland, 2008).

FGF8 is expressed when the zygotic genome is activated at the mid-blastula transition in *Xenopus* and is present in two alternatively spliced forms, FGF8a and FGF8b (Branney et al., 2009; Fletcher et al., 2006). FGF8a was shown to have little mesoderm-inducing activity (Christen and Slack, 1997) and thought to be more important for neural patterning (Hardcastle et al., 2000). By generating an antisense morpholino oligo that specifically targeted FGF8a, Fletcher et al. (2006) showed that FGF8a is essential for posterior neural cell fate and that FGF8b is the major isoform important for mesoderm formation.

Another important transcriptional target of FGF signalling is *Egr1* (early growth response 1) (Branney et al., 2009) that codes for a zinc finger transcription factor and, like *Xbra*, is expressed throughout the marginal zone (Panitz et al., 1998). A detailed study into the regulation of *Egr1* has shown that the ETS-box transcription factor Elk-1 is phosphorylated by MAPK and interacts with SRF to form a complex that binds to regulatory sequences upstream of *Egr1* and, in this way, FGF signalling directly activates the transcription of *Egr1* (Nentwich et al., 2009). Egr1 itself represses the transcription of *Xbra*, but activates the expression of *MyoD* (Nentwich et al., 2009), another immediate early target of FGF signalling (Fisher, 2002). Nentwich et al. (2009) demonstrate that this effector of FGF signalling, Egr1, can act to promote transcription of one gene and repress transcription of another gene, both of which are known to be positively regulated by FGF signalling. These data point to the complexity of the gene networks downstream of signalling that manage to orchestrate cell fate decisions during development.

A detailed picture of FGF-dependent mesodermal gene expression is now available. A large-scale analysis of genes normally expressed in early gastrula-stage embryos was compared to those expressed in sibling embryos in which FGF signalling was inhibited using the dominant negative

FGFR (Branney et al., 2009). Another study compared gene expression in control mesoderm versus that in mesoderm in which FGF signalling was inhibited with the FGFR inhibitory drug SU5402 (Chung et al., 2004). These transcriptomic approaches have identified many genes already known to depend on FGF, such as *Xbra* and *Cdx4*, which internally validated the analysis, as well as a large number of new genes including a novel MAP kinase phosphotase (MKP) called *DUSP5* (Branney et al., 2009). It is clear from these analyses that a subset of the genes transcriptionally activated by FGF signalling are those coding for factors known to down-regulate FGF signalling, such as the MKPs and the Sproutys. Transcriptionally activating negative regulators of FGF signalling may be a mechanism to balance the positive feedback loop working through *Xbra*, which is also a transcriptional target of FGF signalling. These findings suggest that both positive and negative feedback loops act to fine-tune the level of FGF signalling during development (Branney et al., 2009).

In addition, Branney et al. (2009) identified a set of genes expressed in the Organiser as targets of FGF signalling, including *noggin*, *chordin*, and *goosecoid* (among others). The notion that FGFs induce Organiser genes is in stark contrast with the original view of FGFs acting as part of the ventral vegetal (VV) signal in the three-signal model (Figure 5). However, in support of the microarray data, these genes were also found to be sensitive to inhibition of FGF signalling in other studies using SU5402 (Delaune et al., 2005; Fletcher and Harland, 2008). In addition, the level of pSMAD1 is significantly increased in embryos in which FGF signalling is blocked; this indicates an increase in BMP signalling in these embryos and is consistent with the down-regulation of genes expressed in the Organiser that code for BMP inhibitors (Branney et al., 2009).

Neural Induction

Spemann's grafting experiments described in the previous section showed that the Organiser, or dorsal mesoderm, can act to induce neighbouring ectoderm to take on a neural fate. Organiser activity is found in the dorsal lip of the blastopore in amphibians, Henson's node in amniotes, and in the embryonic shield in fish. Identifying the signals that underlie this remarkable activity was a major thrust of many labs over many years. The first neural inducer identified was *Noggin* (Smith and Harland, 1992), followed a few years later by *Chordin* (Sasai et al., 1995). These genes are expressed in the Organiser and when overexpressed drive animal cap explants to form neural tissue.

Chordin and Noggin are both secreted BMP antagonists that bind BMP ligands extracellularly and prevent their association with receptors, thereby blocking BMP signalling (Piccolo et al., 1996; Zimmerman et al., 1996). Interestingly, it was known that dissociated ectoderm had a tendency to neuralise (Godsave and Slack, 1989; Saint-Jeannet et al., 1990); that is, just shaking ectodermal explants into single cell suspension with no added factors will result in neural cell differentiation. This observation could now be explained by a model where cell–cell signals in dissociated ectoderm are massively diluted into the surrounding media and it is the inhibition of BMP signalling in these cells that results in neural induction. A "default" model of neural induction was proposed where ectoderm has an autonomous ability to form neural tissue, and this is blocked by BMP signalling (Hemmati-Brivanlou and Melton, 1997). Conclusive, in vivo evidence that BMP inhibitors are essential for neural induction was provided by experiments in *X. tropicalis* that used antisense morpholino oligos (AMOs) to deplete *Chordin*, *Noggin*, and another BMP inhibitor called *Follistatin*; this triple knockdown resulted in a complete loss of the neural plate (Khokha et al., 2005). Therefore, in *Xenopus*, it is clear that the attenuation of BMP signalling is essential for neural induction; these findings fit well with the specific expression of *Noggin* and *Chordin* in the Organiser, and the exclusion of BMP4 expression in cells fated to become neural.

This textbook model of neural cell fate specification has been largely worked out using *Xenopus*, and this model may not fit as well for all vertebrates. In amniotes, the expression of BMP ligands and antagonists is not entirely consistent with this model (Streit et al., 1998) and blocking BMP signalling cell autonomously by electroporating SMAD6 is not sufficient to induce Sox3 expression in competent chick epiblasts (Linker and Stern, 2004). However, it could be that these reagents are just less efficient at inhibiting BMP signalling. A more effective inhibitor is a dominant negative form of SMAD5, based on the zebrafish mutant *somitabun,* that is thought to form nonfunctional multimeric complexes with SMAD1, 5, and 8 to shut down the BMP signalling pathway (Hild et al., 1999). Overexpressing this mutant SMAD5 in ventral epidermis was capable of inducing neural tissue cell autonomously in early neurula frog embryos; an activity not seen when mRNA coding for SMAD6 or a dominant negative BMP receptor is injected into the same cells (Delaune et al., 2005; Linker and Stern, 2004). It is possible that this powerful method of blocking BMP signalling could also result in neural induction in chick epidermis.

It has been shown that chick epiblast explants overexpressing BMP4 are compromised in neural induction assays (Wilson et al., 2000) and that BMP4 overexpression in the chick neural plate inhibits the expression of Sox2, but not the earlier preneural gene Sox3. This suggests that BMP inhibition could play a later role in neural development in chick, but an earlier signal was required to initiate the process (Linker and Stern, 2004). This other signal may be provided by FGF: In chicks, blocking FGF signalling can lead to a loss of neural markers like *Sox3* (Streit et al., 2000), and with experiments in the urochordate *Ciona intestinalis,* FGF appears to be an important neural inducing signal, indicating an ancient, conserved role for FGF signalling in the chordate lineage (Bertrand et al., 2003). In *Xenopus,* blocking the BMP pathway results in the expression of *FGF4,* supporting the notion that both FGF signalling and attentuation of BMP signals are essential for neural induction (Marchal et al., 2009).

One way by which FGF signalling may be acting to induce neural fates is through its ability to inhibit the expression of BMP ligands. In chick, it was found that while chick epiblast explants express *BMP4* and *BMP7* very early on, the expression of these ligands decreased as cells became specified into the neural cell lineage, concomitant with the initiation of *FGF3* expression (Wilson et al., 2000). Blocking FGF signalling with SU5402 was found to inhibit neural specification in these explants, indicating FGF is critical for neural induction in chicks. In zebrafish, FGF was found to restrict the expression of the genes coding for BMP2 and BMP7 to the ventral domain of the blastula embryo. This work also showed that activation of FGF signalling caused a dorsalisation of the embryo by inhibiting BMP gene expression and blocking FGF signalling expanded the expression of BMP ligands and therefore ventralise embryos (Furthauer et al., 2004). This means that work in both chick and fish indicate that FGF signalling is important very early on in setting up the dorsal–ventral axis and restricting the expression of BMP ligands. In addition, in frogs, FGF signalling has been found to be required for the Organiser expression of BMP antagonists such as *Chordin* and *Noggin* (Branney et al., 2009; Delaune et al., 2005; Fletcher and Harland, 2008) as discussed in the section on Mesoderm Induction. These findings point to two separate ways by which transcriptional responses to FGF can influence the BMP signalling pathway to possibly impact neural induction (see Figure 7). Another mechanism for FGF input into neural induction is

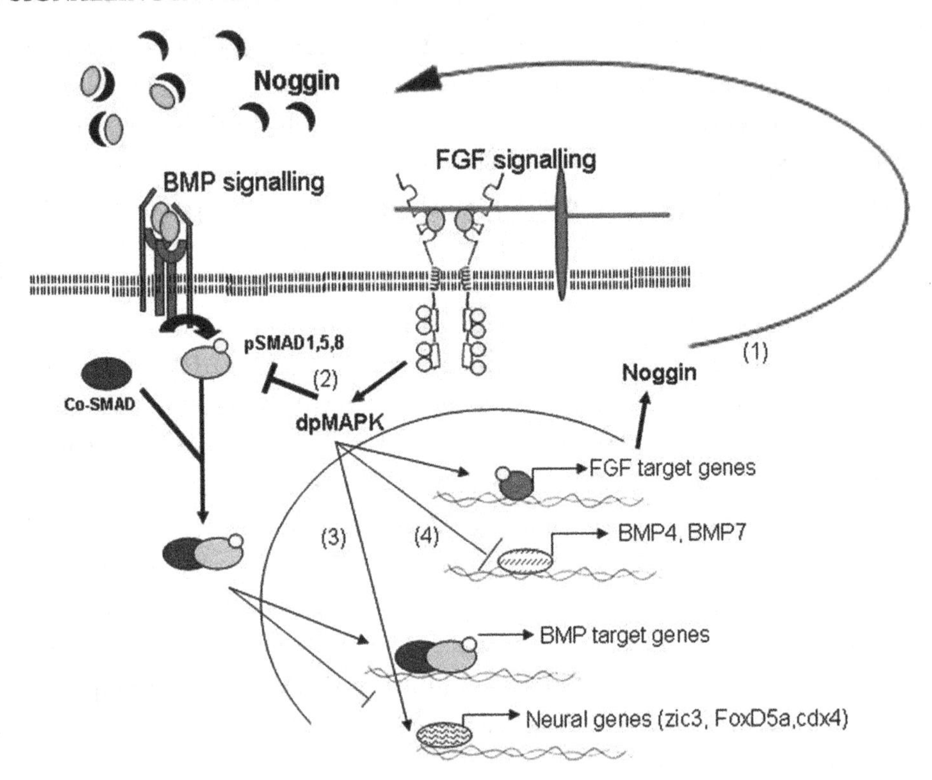

FIGURE 7: Neural induction by BMP and FGF signalling. There are many levels where FGF can impact on neural induction. (1) FGF signalling is required for the expression of Noggin, which acts outside the cell to bind and inhibit the activity of BMP ligands. (2) FGF signalling results in the phosphorylation of SMAD1, 5, 8 in a central domain, which inhibits its ability to move to the nucleus or activate the transcription of BMP target genes. (3) FGF signalling can directly activate the transcription of a set of posterior neural genes. (4) FGF can inhibit the expression of genes coding for BMP ligands.

through the direct modification of the transcriptional effector of BMP signalling, SMAD1. Activation of the BMP receptor (a serine/threonine kinase) results in the phosphorylation of SMAD1 on serine residues at the carboxy-terminus, which causes it to move to the nucleus to activate gene transcription together with a co-SMAD. In contrast, the central domain of the SMAD1 protein can be phosphorylated by MAPK, which results in cytoplasm retention of SMAD1 and inhibits transcription. When the consensus sites for MAPK phosphorylation in the central domain are mutated, injection of mRNA coding for the mutant construct causes ventralisation of embryos and a loss of neural markers, consistent with an overactive BMP signalling pathway (Pera et al., 2003). Interestingly, it has been shown that dissociating animal cap cells, which neuralises the cells, results in the activation of MAPK (Kuroda et al., 2005). These authors propose a molecular mechanism to explain this neuralisation where sustained MAPK activity in the dissociated cells results in phosphorylation of the SMAD1 central domain. This inhibits BMP signalling to promote neural fates. This model suggests that in the embryo, FGF acts through MAPK to inhibit SMAD1 activity and contributes to neural induction.

Other work supports a different mechanism where BMP inhibition and FGF activity act independently during neural induction. Using a combination of knocking down BMP inhibitors in *X. tropicalis* and treating embryos with SU5402 to inhibit FGF signalling, Wills et al. (2010) recently showed that anterior neural markers are more sensitive to depletion of BMP inhibitors, while posterior neural markers were more sensitive to FGF inhibition (Wills et al., 2010). Distinct genetic targets of BMP and FGF during neural induction have been identified. These include *zic3* and *FoxD5a* as two genes directly activated by FGF, and *zic1* as directly activated in response to BMP signalling. These findings suggest that these signalling pathways work independently during neural patterning (Marchal et al., 2009) and are consistent with the observations of Wills et al. (2010) that particular sets of genes are more sensitive to blocking either FGF or BMP signals.

FGF Signalling and Posterior Neural Patterning

The finding that FGF regulates distinct genes from BMP during neural induction and patterning is consistent with many studies that have demonstrated a central and essential role for FGF in posterior development. A classic set of grafting experiments carried out by Peter Nieuwkoop revealed the presence of a posterior localized "transforming" signal. In these experiments, folds of competent ectoderm were grafted to different locations along the anteroposterior axis of an amphibian neurula-stage embryo. The grafts placed closer to the posterior end of the embryo gave rise to more posterior neural cell fates. It is likely that the high level of FGF signalling in the posterior mesoderm contributes to this transforming activity. Explants compromised in FGF signalling do not express posterior neural markers in response to neural induction by Noggin, while anterior markers are unaffected (Ribisi et al., 2000). In addition, overexpression studies driving FGF4 expression during and after gastrula stages dramatically inhibits anterior development and up-regulates the expression of the posterior Hox genes *HoxB9* and *HoxA7* (Isaacs et al., 1994; Pownall et al., 1996). The posterior *Hox* genes as well as the *caudal* homologues Cdx1, Cdx2, and Cdx4 were found to be targets of FGF signalling (Keenan et al., 2006; Northrop and Kimelman, 1994). These genes are key regulators of posterior development and their normal expression requires FGF signalling (Faas and Isaacs, 2009; Isaacs et al., 1994; Isaacs et al., 1998; Pownall et al., 1998). This essential role for FGF in posterior development has been found to be conserved in mice and zebrafish (Ota et al., 2009; Rossant et al., 1997; Shimizu et al., 2006). A role for FGF signalling in specifying neurons in the primary nervous system of amphibians is also suggested by the observation that activation of FGF8 signalling leads to development of ectopic neurons in *Xenopus* (Hardcastle et al., 2000).

In chick, cells in the posterior are proliferative and this is fundamentally linked to the extending anteroposterior. FGF8 is highly expressed in this posterior mesoderm and is thought to maintain a "stem zone" where cells are maintained in an undifferentiated state. As the body axis extends, cells leave the stem zone and down-regulate FGF signalling in the "transition zone" where there are high levels of retinoic acid (RA). FGF and RA have an antagonistic relationship, where RA down-regulates the expression of *FGF8* and FGF signalling inhibits the expression of *Raldh2*, which codes for an enzyme essential for RA synthesis. The posterior activity of FGF induces posterior Hox genes, assigning the anteroposterior character of the neural cells before they leave the "stem zone" (Diez del Corral et al., 2002; Diez del Corral et al., 2003; Diez del Corral and Storey, 2004). In addition, FGF acts in part through the regulation of posterior Wnt signals that promote cells moving through the "transition zone" and becoming responsive to RA (Olivera-Martinez and Storey, 2007). These interactions are also critical for the segmentation of the mesoderm during somitogenesis, and this will be discussed later.

FGF Signalling at the Isthmic Organizer

The isthmus is a constriction at the midbrain–hindbrain junction (MHJ) that acts as an important signalling centre during vertebrate development. This was demonstrated in an elegant experiment by Martinez et al. (1991) when they grafted an MHJ into the posterior forebrain and found that this induced the formation of ectopic midbrain (Martinez et al., 1991). The induced midbrain formed as a mirror image of the endogenous midbrain, suggesting that there is a high concentration of the inducing signal in the graft that diffuses across the tissue in which it is grafted to specify different fates depending on its local concentration. Becaue of this characteristic activity, the MHJ is referred to as the ismthic organizer (IsO).

FGF8 and other members of the FGF8 subfamily including FGF17 and FGF18 are expressed in the IsO in many vertebrate embryos (Figure 8). Many genes that code for elements of the FGF signal transduction pathway, such as FGF-R1 and Sprouty, as well as targets of FGF signalling like Ets transcription factor genes are also expressed in the IsO. Furthermore, it has been shown that FGF8 can recapitulate the organiser activity of MHJ grafting: Implanting an FGF8-soaked bead into the posterior diencephalon transforms it into midbrain that has a mirror-image orientation relative to the endogenous midbrain (Crossley et al., 1996). The expression of *En2*, *Wnt1*, and *FGF8* are induced by the FGF bead, suggesting that the FGF bead is capable of inducing a new signalling centre and that this ability reflects an important role for FGF in the IsO.

The expression of FGF8 in the IsO requires the activity of the homeobox transcription factors Gbx2 and Otx2. Early in development, Otx2 defines anterior neuroectoderm, while Gbx2 defines posterior neuroectoderm. Otx2 is required for the formation of the forebrain and midbrain (Acampora et al., 1998) and Gbx2 is required for the formation of rhombomeres 1–3 (Wassarman et al., 1997). Manipulating the expression of Gbx2 or Otx2 will shift the expression of FGF8 posteriorly or anteriorly, consistent with an antagonistic relationship of these transcription factors defining the MHJ.

The IsO lies at the boundary of the midbrain (the mesencephalon) and the most anterior part of the hindbrain, rhombomere 1. The mesencephalon will go onto form the tectum where auditory

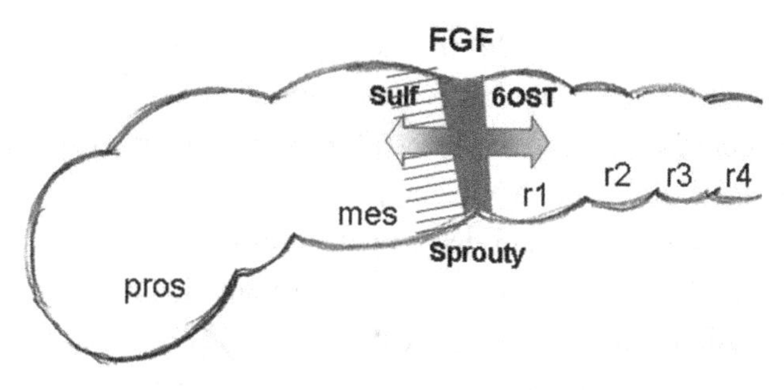

FIGURE 8: FGF signalling at the MHJ acts as an organiser. Higher levels of FGF signalling occur posterior to the MHJ source of FGF and specify the metencephalon (r1). Lower FGF activity anterior to the MHJ specifies the mesencephalon (mes). The mechanism by which less FGF is perceived anterior to the source may involve the absence of appropriate HSPGs (blue stripes) as a region just anterior to the MHJ expresses Sulf2 and does not express 6-*O*-sulfotransferase (6OST). The presence of Sprouty may also regulate the response to FGF in this region. Prosencephalon (pros), mesencephalon (mes), rhombomeres 1–4 (r1, r2, r3, r4).

and visual cues are processed, while rhombomere 1 is part of the metencephalon which will give rise to the cerebellum which coordinates motor activity. A signalling centre at the boundary of these two regions of the embryonic brain may provide a mechanism to allow the coordinated development of the tectum and the cerebellum that will need to functionally interact in the fully developed organism. The Hox code is a highly conserved system that gives cells positional information along the anterior posterior axis of the developing central nervous system; however, the most anterior boundary of Hox gene expression is at rhombomere 2. It therefore seems logical to have another means of patterning the tissue in this region of the CNS, and FGF activity in the IsO is thought to play this role (reviewed by Mason, 2007).

The finding that different levels of FGF signalling can elicit a different tissue response suggests a mechanism for organiser activity of the IsO. For instance, FGF8b (which is a strong activator of the MAPK pathway) will convert mesencephalon to rhombomere 1, while FGF8a (a weaker activator of MAPK signalling) transforms posterior forebrain to mesencephalon (Sato et al., 2004; Sato and Nakamura, 2004; Suzuki-Hirano et al., 2005). Note that both of these activities are transforming an anterior fate to a posterior one, consistent with a role for FGF as a posteriorising factor. More recently, a conditional knockout using an En1-Cre mouse to target a specific mutation in FGF8b to the IsO has shown that the expression of FGF8b is essential for the expression of all IsO genes analysed (Guo et al., 2010), while FGF8a was not. One model suggests that lower FGF activity specifies mesencephalon (which lies anterior to the IsO) and higher FGF activity is needed to specify rhombomere 1 (which lies posterior to the IsO) (Basson et al., 2008). It is a puzzle how

FGF ligand, which is present at the MHJ, can be perceived differently in the cells lying posterior and anterior to the same source; this will be addressed below and in Figure 8.

A different kind of approach to detect FGF ligands in vivo has been taken recently: Chen et al. (2009) used FGF-R3 fused to alkaline phosphatase to map levels of FGF across the chick MHJ (Chen et al., 2009). They found that a gradient of FGF was present with higher levels present in the posterior midbrain and lower levels in the anterior midbrain. Midbrain explants also showed distinct responses to different concentrations of FGF, where high concentrations induce the expression of ephrin A and low concentrations induced the expression of the EphA receptor. In this case, the posterior-to-anterior gradient in the mesencephalon produces a graded output of gene expression. These authors suggest that the high levels of FGF in the hindbrain compared to the graded FGF in the midbrain would result in the discrete output of rhombomere 1 genes versus tectal (mesencephalon) genes.

Another finding in this paper was that the gradient of FGF across the IsO requires the presence of HSPGs (Chen et al., 2009). As discussed above, HSPGs are known to be required for the formation of the tripartite signalling complex that is comprised of FGF peptides, FGFRs, and heparan sulphate chains (Turnbull et al., 2001). It is further known that the HS must be modified by 6-O-sulfation to interact with this complex (Pye et al., 2000) and that the enzymes Sulf1 and Sulf2 can disrupt this complex and inhibit FGF signalling by specifically removing these moieties (Wang et al., 2004). In light of this, it is interesting to observe that in *Xenopus*, there is a stripe of Sulf2 expression in the posterior mesencephalon as well as an overlapping gap in the expression of 6-*O*-sulfotransferase (Winterbottom and Pownall, 2009), the enzyme required to add the 6-*O*-sulphate group to HS during its biosynthesis. This expression pattern may reveal one mechanism that allows the region just anterior to the MHJ to be less responsive to FGF and the hindbrain to perceive more FGF. The local structure of HSPGs is less conducive to FGF signalling in the mesencephalon due to the presence of Sulf2 (see Figure 8). Another negative regulator of FGF signalling, *Sprouty2*, is also expressed in the IsO and its expression is known to be rapidly activated by FGF8. Electroporation of *Sprouty2* into chick rhombomere 1 converts it to mesencephalon (Suzuki-Hirano et al., 2005), consistent with the idea that a lower level of FGF signalling is required to specify the midbrain.

FGF Signalling at the Anterior Neural Ridge

The anterior neural ridge (ANR) lies where the most anterior neural tissue meets the nonneural ectoderm forming a smile-shaped junction that can be visualised by looking at the expression of FGF8 (Figure 9). The ANR has been shown to be essential for the formation of the telencephalon because removal of this structure in mice or fish will result in the lack of forebrain expression of the key telencephalon marker and regulator, the forkhead transcription factor *Foxg1* (Houart et al., 1998). In addition, similar to the experiments described above for the IsO, grafting an ectopic ANR induces the expression of telencephalic markers in the diencephalon, which indicates that this region may also act as an organising centre to pattern the forebrain. Furthermore, an FGF bead implanted into embryos that lack the ANR can compensate and induce *Foxg1* expression (Shimamura and Rubenstein, 1997). So again, FGF signalling has been found to play a central role in mediating the activity of another signalling centre in the developing brain, the ANR.

Several FGF ligands have been found to be expressed in the ANR, including *FGF8* and *FGF3*, as well as three of the four genes that code for the FGFRs (*FGF-R1*, *R2*, and *R3*). This overlapping expression of many receptors and ligands has resulted in a lack of phenoptype in the telecephalon of mice or fish mutant for any one of these players due to the presence of related genes that can compensate for its loss. However, when the genes coding for FGF R1–3 are deleted from the telecephalon in the mouse, the telencephalic markers *foxg1*, *dlx2*, and emx1 are not expressed and cells in this region die (Paek et al., 2009). This work clearly demonstrates the essential role for FGF signalling in the induction and survival of the telecephalon. In addition, *FGF8* expression in the ANR requires Foxg1 (Martynoga et al., 2005), suggesting that these genes are co-regulated in a positive feedback loop. Interestingly, FGF signalling has also been shown to regulate intracellular localisation of Foxg1 protein. FGF signalling working through an AKT-dependent pathway promotes nuclear export of Foxg1 and neuronal differentiation in the forebrain (Regad et al., 2007).

Shh is present in the ventral telencephalon and is known also to be required for *Foxg1* expression. Shh promotes ventral cell fates in the forebrain by antagonising the dorsally active Gli3 (Hebert and Fishell, 2008). *FGF8* expression in the ANR was found to be regulated by Shh, as its

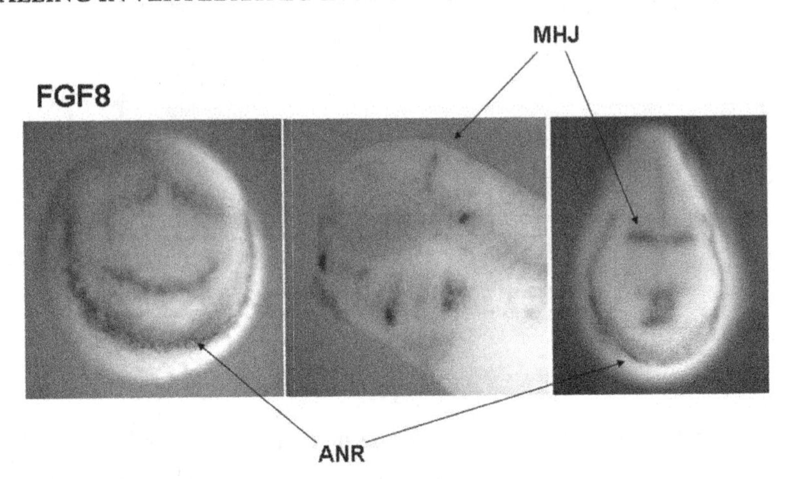

FIGURE 9: FGF8 expression in the ANR and the MHJ in *X. tropicalis*. The panels on the right and the left show a face-on (anterior) view, while the middle panel has anterior to the left. All are shown dorsal side up.

expression is up-regulated when both *Shh* and *Gli3* are knocked out. Shh normally represses Gli3 activity, which in turn normally represses *FGF8* expression; this means that Shh signalling results in the activation of *FGF8* expression. In addition, in the FGFR1–3 knockout mice, *Shh* expression is reduced in the telecephalon at E10.5, suggesting that FGF signalling is required to maintain *Shh* expression (Paek et al., 2009). This type of Shh-FGF positive feedback loop is also seen during limb development and is discussed below.

Morphogenesis

GASTRULATION

In animal embryos, cells not only receive and interpret signals to commit to particular a lineage, but in addition, cells can also change their cytoskeletal structure and behaviour so that, in response to signals, cells move. The process by which cells move during development to change tissue shape and the relative positions of different cell types is known as morphogenesis. It can be a difficult challenge to dissect distinct effects of cell signals in determining cell fate versus their effects on morphogenesis. In *Xenopus*, early canonical Wnt signalling (that acts through the stabilisation of β-catenin) on the dorsal side of the embryo specifies the cells of the organiser (Tao et al., 2005). Subsequently, these cells drive the initiation of gastrulation movements, including a process called convergent extension (CE) (Figure 10). During CE, the coordinated polarisation and intercalation of dorsal mesoderm cells drives the lengthening of the anteroposterior axis. Interestingly, this cell behaviour is known to require gene products in the planar cell polarity (PCP) pathway, one of the noncanonical Wnt signalling pathways (Tada et al., 2002). Thus, both canonical and noncanonical Wnt signalling pathways are central in the specification and behaviour of organiser cells.

In amniotes, a different cell behaviour underlies gastrulation: as cells enter the primitive streak, they undergo a epithelial to mesenchymal transformation (EMT); this is where a cell that is part of an epithelium loses contact with its neighbours, changes shape, and moves free. *Snail* codes for a transcriptional repressor that inhibits the expression of *E-cadherin*, a gene that codes for a cell adhesion protein. Activation of *Snail* down-regulates *E-cadherin*, which loosens cell–cell association in an epithelium so that a cell can migrate away. Importantly, this kind of cell behaviour is also characteristic of metastatic cancer cells and E-cadherin is known to act as a tumour suppressor (Jeanes et al., 2008). Analysis of chimeric mice mutant for FGFR1 has shown that cells require FGF signalling to down-regulate *E-cadherin* and undergo EMT; cells that fail to do this are trapped in the primitive streak (Ciruna and Rossant, 2001; Ciruna et al., 1997).

There are severe gastrulation defects in amniote and fish and frog embryos deficient in FGF signalling, including FGFR1 mutant mice (Ciruna and Rossant, 2001) and zebrafish and *Xenopus* embryos injected with mRNA coding for a dominant negative form of FGFR 1 (Amaya et al., 1991; Griffin et al., 1995; Isaacs et al., 1994). These data point to an important role for FGF signalling in

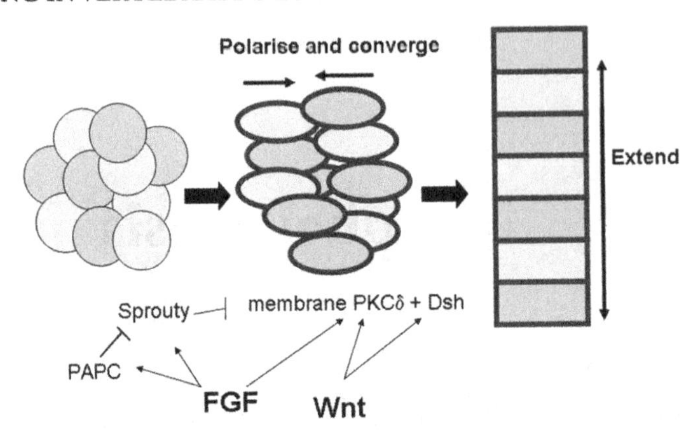

FIGURE 10: Convergent extension (CE). Activation of the planar cell polarity pathway results in the membrane localisation of dishevelled (Dsh) and PKCδ and the reorganisation of the cytoskeleton such that cells become polarised along on axis. Polarised cells intercalate, which causes the tissue to extend along the perpendicular axis. FGF and noncanonical Wnt signalling can activate the PCP pathway and are required for normal CE movements.

regulating the diverse strategies for gastrulation in different vertebrates. In chicks, an in vivo study showed that cells migrating through the primitive streak during gastrulation will take different paths depending on their proximity to sources of FGF ligand. FGF4 beads implanted into gastrula-stage chick embryos were found to act as a chemoattractant and GFP-expressing mesodermal cells migrated towards the FGF4 beads. In contrast, FGF8 beads repelled labelled mesodermal cells. The authors' interpretation of these findings are that cells close to the chick organiser (Hensen's node) are attracted to the FGF4 present in the node and early notochord and so they migrate anteriorly, while FGF8, which is expressed more posteriorly in the late primitive streak, acts as a repulsive cue encouraging cells to migrate away from the streak (Yang et al., 2002).

During and subsequent to gastrulation, the anteroposterior axis of chordates lengthens by means of CE. The PCP pathway acts intracellularly to polarise a cell by driving proteins including Dishevelled (Dsh) to move to a particular part of the cell membrane. However, CE requires a field of cells to act together and to coordinately polarise and intercalate with each other. In *C. intestinalis*, *FGF3* is expressed in the ventral midline of the nerve cord (neural tube), lying just dorsal to the forming notochord that expresses FGFR. Notochord cells expressing dnFGFR, or depleted of FGF3, fail to intercalate suggesting that FGF signalling provides an exogenous cue that gives directionality to the intracellular PCP pathway (Shi et al., 2009). Interestingly, dpERK is not active in the developing notochord in *Ciona*, indicating that the MAPK pathway is not important in this process. The dorsal mesoderm in *Xenopus* embryos overexpressing dnFGFR1 also fails to extend along the anteroposterior axis and the abnormal cell movements exhibited by the dorsal mesoderm in these embryos results in their characteristic open-blastopore phenotype (Isaacs et al., 1994). These data clearly demonstrate that the cells of the dorsal mesoderm, that is, the organiser, require FGF signalling for appropriate cell movement.

Sprouty was identified in *Drosophila* as an antagonist of FGF signalling (Hacohen et al., 1998) and in *Xenopus*, Xsprouty2 has been shown to inhibit morphogenesis while not effecting mesoderm specification (Nutt et al., 2001). A related protein called Spred inhibits MAPK activation by FGF signalling, so it has been suggested that these two proteins act on different intracellular pathways downstream of FGF (Sivak et al., 2005). This conclusion was based on the finding that Sprouty has inhibited FGF-mediated calcium release, membrane localisation of PKCδ, and morphogenetic movements, without affecting dpERK activity or the expression of *Xbra*. In contrast, Spred did not inhibit membrane localisation of PKCδ, but did abolish *Xbra* expression and the activation of dpERK. Another study has shown that Sprouty, and not Spred, physically associates with paraxial protocadherin (PAPC) (Wang et al., 2008), a protein known to be important for CE movements in fish and frogs (Kim et al., 1998; Yamamoto et al., 1998). Because Sprouty is known to inhibit PCP-directed morphogenesis, and PAPC is known to be a positive regulator of PCP, one model is that the physical association of PAPC and Sprouty weakens Sprouty's inhibitory activity (Wang et al., 2008). This notion is supported by experiments that show membrane localisation of PKCδ, stimulated by expressing the Wnt receptor Frizzled-7 (Fz7), is inhibited by Sprouty and that this inhibition is prevented by PAPC. PAPC on its own does not promote membrane recruitment of PKCδ indicating that PAPC is acting by sequestering Sprouty. Interestingly, the expression of both *Sprouty* and *PAPC* in early *Xenopus* embryos requires FGF signalling (Branney et al., 2009; Nutt et al., 2001). The neutrophin receptor homologue is a known transcriptional target of FGF signalling, which is required for normal CE and provides another plausible node of interaction for the FGF and PCP pathways (Chung et al., 2005). These data weave together the FGF signalling pathway with the PCP pathway downstream of noncanonical Wnt signalling.

INNER EAR

Studies in amniote embryos have shown that FGF signalling is required for the development of the inner ear (Ladher et al., 2005; Wright and Mansour, 2003). FGF has been found to play a role in the induction of the otic vesicle and the activation of genes required for inner ear development including *Pax2* and *Nkx5.1*. In addition, FGF also drives some of the morphological changes underlying ear development. Early FGF signalling leads to the formation of a progenitor domain that will give rise both to the inner ear and to the epibranchial placode from which several different ganglia will eventually arise. Mesodermally expressed FGF3 and FGF 19 promote proliferation in an otic-epibranchial precursor domain, and it is the attenuation of FGF signalling together with canonical

Wnt signals that specifies inner ear fate, while continued FGF signalling promotes epibranchial placode development (Freter et al., 2008). After the otic placode is induced, the ectoderm thickens and subsequently invaginates to form a hallow ball called the otocyst that lies within the head mesenchyme. Explant studies in chick showed that invagination of isolated otic ectoderm requires application of FGF to the basal side of the tissue (Sai and Ladher, 2008). FGF was found to induce basally localised PLCγ and myosin II, which causes a local depletion of actin fibres on the basal part of the cells and an enrichment of actin filaments apically. These effects of FGF were found not to be sensitive to protein synthesis inhibitors, indicating that in these cells, FGF can directly remodel the cytoskeleton to influence morphogenesis of a tissue. Later during ear development, a specialised region of the inner epithelium called the organ of Corti is patterned into a highly ordered array of sensory and nonsensory cells. The nonsensory cells are supporting cells, and the sensory cells are hair cells that act as mechanosensory receptors and are critical in translating the sound waves into electrical signals that are sent to the brain. In addition to the well-described role of Notch signalling in patterning inner ear development (Daudet and Lewis, 2005), FGFRs 1 and 3 have been found to be important in specifying cell types in the auditory sensory epithelium (Colvin et al., 1996; Pirvola et al., 2002). The precise level of FGF8 activity in the organ of Corti is modulated by the presence of Sprouty2, and the amount of FGF signalling is interpreted to produce support cells called pillar cells. In the absence of Sprouty2, higher levels of FGF result in three pillar cells instead of two forming, which leads to a deformed cochlea and a deaf mouse. These effects can be partially rescued by genetically reducing FGF8 levels in *Sprouty2* mutant mice (Shim et al., 2005). Hair cells are marked by the expression of the proneural gene *atonal homologue 1* (*Atoh1*), and treatment with the FGF inhibitor SU4502 in zebrafish has shown that the expression of atoh1 and hair cell development requires continuing FGF signalling (Millimaki et al., 2007).

LATERAL LINE

Mechanosensory hair cells are also found in another organ that requires FGF during its morphogenesis. In zebrafish, the lateral line placode is a migrating epithelium that deposits precursors of hair cell organs, called neuromasts, at regular intervals in a line along each side of the embryonic trunk. As the lateral line placode migrates from head to tail, rosettes of cells bud off from the epithelium just behind the leading edge. These rosettes, or proneuromasts, will give rise to the neuromasts. The leading edge of this migrating epithelium undergoes a partial EMT, or a pseudo-EMT, where cells at the leading edge lose some apicobasal polarity and have increased numbers of filapodia but remain in contact with neighbouring cells. FGFR1, FGF3, and FGF10 are all expressed in the migrating lateral line primordium, where the ligands are specifically expressed in cells at the leading edge and the receptor in the trailing region. In zebrafish embryos mutant for FGF3 and FGF10, or in embryos treated with the FGF inhibitor SU5402, migration of the lateral line primordium ceases and rosettes are not formed (Lecaudey et al., 2008; Nechiporuk and Raible, 2008). Time lapse microscopy revealed a requirement for FGF3 and FGF10 for the formation of the proneuromasts and that the formation of these rosettes is a prerequisite for the migration of the lateral line primordial (Nechiporuk and Raible, 2008). Together, the findings point to a role for FGF signalling in regulating the pseudo-EMT that underlies the deposition of neuromasts along the trunk.

PRONEPHROS

FGF plays a role in EMT during amniote gastrulation and lateral line migration in zebrafish; however, FGF is also known to be essential for mesenchymal to epithelial transformation (MET) during the morphogenesis of the *Xenopus* larval kidney, the pronephros. The pronephros is comparable physiologically, morphologically, and developmentally to the cortical nephron present in the adult kidney of the frog and the metanephric kidney in mammals. The pronephric mesoderm is derived from the intermediate mesoderm and the condensation of the pronephric mesenchyme can be inhibited with SU5402. Pronephric morphogenesis begins as this mesenchyme segregates away from the lateral plate and somites and begins to migrate posteriorly. The pronephric duct forms as the migrating pronephric mesenchyme undergo epithelialisation in an anterior to posterior wave to form a hollow tube that fuses with the cloaca. *FGF8* is expressed transiently in the *Xenopus* pronephros during its morphogenesis, and morpholino knockdown of FGF8 was found to block the development of the pronephric duct (Urban et al., 2006). A key aspect of these effects was the loss of epithelialisation in the pronephric duct in embryos lacking FGF8. There are therefore two stages of *Xenopus* kidney development that require FGF signalling: the specification and condensation of the pronephric mesoderm and its epithelialisation to form the pronephric duct. Interestingly, FGF signalling is also known to be important in the branching, cell proliferation, and growth of the mammalian metanephric kidney (Qiao et al., 2001). A recent study indicates that FGFR1 is required for the development of the mouse metanephric kidney (Gerber et al., 2009).

BRANCHING MORPHOGENESIS

There are branched tubular networks in many vertebrate organs such as the kidney, the lungs, and the vasculature. Most of these develop from a simple epithelial sac that undergoes reiterative budding to form a complex, tree-like array. A clear and important role for FGF signalling during branching morphogenesis has been established for the development of the trachea system in the *Drosophila* larva (Metzger and Krasnow, 1999). *Branchless* (an FGF ligand) is expressed in cells surrounding the epithelial trachea cells and is responsible for activating *Breathless* (an FGFR) in the trachea cells, which results in the expression or activation of downstream components of the FGF pathway. *Sprouty* was also first identified by its effects on branching morphogenesis in *Drosophila* (Hacohen et al., 1998). Spry is expressed in cells close to the source of FGF and is important for locally preventing secondary branching. Similarly, FGF signalling has been found to be a key regulator of branching morphogenesis in the mouse lung.

The developing mouse lung is composed of an epithelial layer, the surrounding mesenchyme, and the outer mesothelium. Signalling between these layers is critical for the coordinated mesenchymal growth and epithelial branching essential for normal lung development. The branching morphogenesis from which the lung evolves has recently been described in detail and found to be remarkably stereotypical (Metzger et al., 2008). There are several FGF ligands expressed in the developing lung. FGF9 is expressed in the outer mesothelium as well as in the epithelial layer, and in $FGF9^{-/-}$ knockout mouse, there is a loss of mesenchymal proliferation as well as reduced branching (Min et al., 1998), suggesting FGF9 is a part of the epithelial to mesenchyme signal. Using explant cultures of mouse lungs, FGF9 was found to stimulate mesenchyme proliferation when supplied on a heparin bead, while FGF10 did not have this activity. FGF9 and canonical Wnt signals promote the proliferation of lung mesenchyme. These effects could in part be attributed to the ability of epithelial FGF9 to activate mesenchymal expression of *sonic hedgehog* (*Shh*), a signal known to promote proliferation of undifferentiated cells (White et al., 2006; Yin et al., 2008). FGF10 is expressed in the lung mesenchyme and FGFR2 is expressed in the lung epithelium; loss of function of either of these genes results in a total loss of branching (Abler et al., 2009; Arman et al., 1999; Min et al., 1998). FGF10 signals from the mesenchyme act on the epithelium to induce the expression of target genes that include *Spry* and *Shh*. Spry feeds back to inhibit MAPK activity in the tip cells and restrict the formation of new branches (Figure 11); Shh also inhibits local FGF signalling by repressing *FGF10* expression (reviewed by Affolter et al., 2009; Horowitz and Simons, 2008).

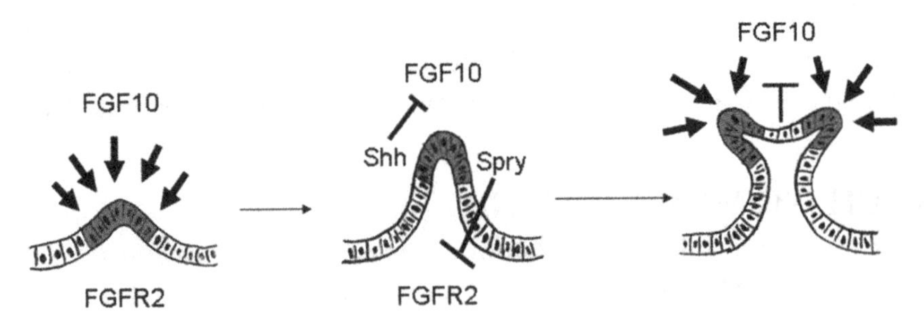

FIGURE 11: FGF10 signals to drive branching morphogenesis in the developing mouse lung. FGF10 is present in the mesenchyme, while FGFR2 is expressed in the epithelium. In response to FGF signals, the epithelium forms a bud and also activates the expression of *Sprouty2* and *Shh*. Both of these gene products dampen the activity of the MAPK pathway; Shh is thought to do this by inhibiting the expression of FGF10. Turning down FGF signalling restricts the formation of buds, in this way other parts of the tubule still competent to respond to FGF will now form new buds.

THE DEVELOPING HEART

In the developing embryo, there are two distinct sources of cells that give rise to the heart: the first heart field (FHF), which forms early in the lateral plate mesoderm to form the cardiac crescent during primitive streak stages, and the secondary heart field (SHF), which is initially located anterior and dorsal to the cardiac crescent in the pharyngeal mesoderm (Buckingham et al., 2005; Srivastava, 2006). The FHF fuses at the midline to form a simple heart tube composed of inner endocardial cells and outer myocardial cells. The morphogenesis of the heart continues as it loops dramatically to the right and the posterior part of the tube shifts towards the anterior. The cells from the SHF remain as progenitors until they migrate and incorporate into the heart, using the FHF as a scaffold. Fate mapping studies in mice using *FGF10:LacZ* transgenics to mark cells in the pharyngeal mesoderm revealed that while SHF gives rise to cells that populate most of the fully developed heart, cells from the FHF are largely restricted to the left ventricle (Zaffran et al., 2004). FGF8 and FGF10 are expressed in the SHF, but only FGF8 has a mutant phenotype, where an FGF8 hypomorph displays septation defects in the outflow tract (OFT) (Abu-Issa et al., 2002). A malformed OFT underlies many congenital heart diseases; failure of OFT septation disrupts the correct partitioning of blood flow because the OFT is not divided into the pulmonary artery and the aorta.

Using conditional loss-of-function mutants for *FGFR1* and *FGFR2* and gain-of-function *Sprouty*, the effects of FGF signalling on OFT morphogenesis was investigated (Park et al., 2008). When these mutant constructs were expressed in the SHF, severe OFT defects were apparent where endothelial cells failed to migrate and undergo EMT, in addition to a fewer number of mesoderm cells in these mutants. In contrast, knocking out FGF signalling in the endothelial cells caused no OFT defects. The requirement for FGF was determined to be in the mesoderm, as endothelial defects could be rescued with grafts of wild-type mesoderm suggesting the effects of blocking FGF signalling was not cell-autonomous. Interestingly, a microarray analysis of OFT derived from FGFR mutant SHF revealed that FGF is required not only for known targets of the FGF pathway, but also for the activity of the BMP pathway. BMP signalling is known to be important for the recruitment and differentiation of myocardial cells during heart development.

In zebrafish, FGF signalling has been found to be important for establishing the correct heart chamber proportions (Marques et al., 2008) and FGF8 mutants have been found to have notably small ventricles (Reifers et al., 2000). Using a drug (SU5402) to inhibit FGF signalling and a zebrafish line transgenic for an inducible activated FGFR, the effects and requirement for FGF signalling in heart development were tested at different developmental time points (Marques et al., 2008). These studies found that FGF plays both an early role in establishing pools of cardiac progenitors and later in regulating the number of cells that give rise to the ventricular myocardium.

Somitogenesis

Segmentation of the vertebrate body is driven by the regular repeated formation of somites from the paraxial presegmental (or presomitic) mesoderm (PSM). Somites are transient mesodermal structures that form as balls of epithelial cells from the anterior part of the PSM and will give rise to the vertebrae and ribs as well as all of the skeletal muscle in the body and the dermis of the skin. In addition to forming the overtly segmented vertebrae, the somites also provide extracellular guidance cues for migrating neural crest and axons, which means that somitogenesis also underpin the segmented pattern of the nervous system.

The process of somitogenesis passes through the embryo in an anterior to posterior wave where somites form one pair at a time with a clock-like rhythm. There is evidence for an internal oscillator that regulates this process and several genes downstream of Notch signalling have been identified as having a rhythmic expression pattern in time with somite formation. In the chick PSM, the expression of the genes *cHairy-1*(*Hes1*) and *lunatic fringe* (among others) cycle in time with the formation of a pair of somites. Related genes are also known to cycle in mouse, fish, and frog embryos indicating that this molecular mechanism is conserved in vertebrates (Dequeant and Pourquie, 2008).

Somite segmentation is coupled to the posterior extension of the body axis. In the posterior of amniote embryos, cells continue to be produced and move through the primitive streak. These cells are present in the posterior PSM as loosely associated mesenchyme and as they move towards the anterior, they organise into an epithelium to form the somitomeres, which are presumptive somites. There is a transition point in the PSM along the anteroposterior axis where cells begin to form somitomeres; this is called the determination front. The determination front is manifest by the expression of a bHLH gene called *Mesp2* (*thylacine*) in the anterior three somitomeres, marking future somite boundaries (Oginuma et al., 2008). The determination front moves in a posterior direction like a wave as the axis of the embryo elongates in an anterior to posterior direction.

A clever microarray analysis has identified FGF signalling as a key component the pathway regulating somitogenesis. In this study, individual PSM explants were excised and processed on chips, after which they were retrospectively assigned to a particular phase of the somitogenic cycle by assessing the expression lunatic fringe in the remaining embryo (Dequeant et al., 2006). Genes downstream of FGF were activated with a distinct expression profile during somite segmentation. *FGF8* is expressed at high levels in the posterior PSM and has been shown to keep the posterior

PSM cells in an immature state (Delfini et al., 2005). Activating FGF signalling by electroporating constitutively active MEK plasmids into chick PSM was used to manipulate the FGF signalling pathway. Using time-lapse video microscopy, it was demonstrated that cells with high levels of MAPK activity maintain high motility that is characteristic of the cells normally present in the posterior PSM, while cells in the anterior PSM normal are less motile as they undergo MET to form a somitomere. The role of FGF in maintaining cells in an immature, undifferentiated state in the tailbud is not restricted to the PSM cells, as neural precursor have also been found to require a reduction in FGF signalling before they will differentiate (Diez del Corral et al., 2002). Ectopic FGF ligand also interferes with segmentation (Dubrulle et al., 2001) as demonstrated when PSM exposed to a bead soaked with FGF8 gives rise to irregular-sized somites. Furthermore, conditional knockouts of FGF-R1 in the posterior mesoderm results in mice with large, irregular somites that give rise to fused vertebrae and irregular vertebrae and ribs (Wahl et al., 2007).

A working model to explain these results is illustrated in Figure 12. High levels of posterior FGF decreases anteriorly, creating a gradient of FGF activity in the PSM. Below a critical threshold level of FGF signalling is the determination front marked by the expression of *Mesp2* and the

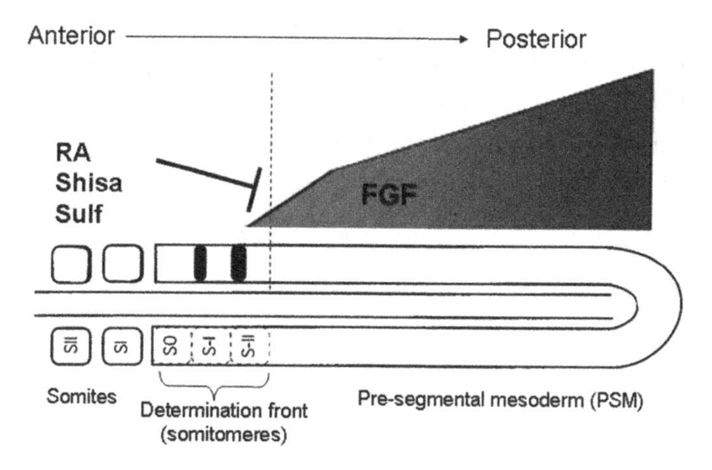

FIGURE 12: A posterior to anterior gradient of FGF establishes the determination front. There is a high level of FGF signalling in the posterior mesoderm; this is opposed by RA signalling that is present in the somites. FGF promotes the motile undifferentiated nature of the cells in the PSM, and only when levels of FGF have dropped below a certain threshold will cells undergo MET to form somitomeres. This threshold defines the determination front, where somitomeres take shape and are marked by the expression of Mesp2 (as black stripes). The activity of RA, and possibly other negative regulators of FGF signalling such as Shisa and Sulf1, represses FGF signalling such that it drops below the required threshold (dotted line) allowing the segmentation of a pair of somites. Also critical, but not shown, are the activities of the cyclic genes that allow cells to respond to the determination front and Wnt signals that act in the posterior with FGF (see Dequeant and Pourquie, 2008).

formation of somitomeres. One important aspect of this model is that posterior FGF is opposed by the anterior signal RA. Indeed, FGF and RA have been found to be mutually inhibitory in both the PSM (Moreno and Kintner, 2004) and the neuroepithelium (Diez del Corral et al., 2003). When the FGF pathway is activated by constitutively active MEK, the expression of the RA-synthesising enzyme *raldh* is inhibited. Moreover, the expression of *cyp26* (a P450 enzyme that metabolizes RA) is ablated in embryos lacking FGF. These data indicate that FGF signalling represses *raldh* expression and is required for *cyp26* expression and this is how FGF negatively regulates RA biosynthesis. Experiments treating embryos with chemical inhibitors of FGF furthermore support the notion that RA represses FGF. In this way, the positioning of somite boundaries at the determination front is established (Moreno and Kintner, 2004). More evidence that an FGF gradient is important for determining where the transition from PSM to segmented somite occurs comes from studies on two cell-autonomous inhibitors of FGF signalling, Shisa2 (Nagano et al., 2006) and Sulf1 (Freeman et al., 2008). The genes that code for both of these regulators are expressed in the paraxial mesoderm anterior to *FGF8* expression. Knockdown of either Shisa2 or Sulf1 in *Xenopus* results in an anterior shift of dpERK activity and a corresponding anterior shift in *Mesp2* (*thylacine*) expression. Shisa acts to trap the FGFR in the endoplasmic reticulum (Yamamoto et al., 2005), while Sulf1 discourages the formation of the FGF signalling complex at the cell surface (Wang et al., 2004). These studies show that such factors within cells reduce the level of signal that is perceived, lowering the effective level of FGF signalling at the determination front. These factors contribute to the local level of FGF activation, and it has been shown that interference with these regulators changes where somites will form (Freeman et al., 2008; Nagano et al., 2006).

FGF and Myogenesis

Skeletal muscle is derived from precursor cells called myoblasts that have been studied extensively in tissue culture over many years (Konigsberg, 1963). Myoblasts taken from embryos replicate clonally in culture and when the amount of growth factors in the media is reduced they differentiate. To do this, myoblasts (1) stop dividing, (2) express contractile protein genes (such as myosin, actin, and troponin, etc.), and (3) undergo cell–cell fusion to form multinucleated myofibres. The key growth factor repressing myogenic differentiation in these cultures was found to be FGF (Clegg et al., 1987). However, distinct clones of primary myoblasts were found to respond differently to FGF treatment: in some, FGF delayed differentiation, as predicted, while in other cases, FGF was found to be required for differentiation (Seed and Hauschka, 1988). In addition to these effects of FGF on cultured muscle cells, the expression of FGF ligands and receptors in the developing muscle of vertebrate embryos suggests that FGF signalling plays a role during skeletal myogenesis in vivo.

As already discussed, somites form sequentially from the posterior mesoderm adjacent to the neural tube and notochord. All skeletal muscles in the body originate from the somites and the earliest cells committed to the myogenic lineage are found in the medial part of the newly formed somite. These cells are discernible by their expression of one or more of the genes coding for the myogenic regulatory factors (MRFs). The MRFs, which include the genes *myoD* and *myf5*, are not only the earliest markers of skeletal muscle progenitors but are also essential regulators initiating the myogenic programme; they code for basic helix–loop–helix transcription factors that act as a developmental switch to drive a cell down the myogenic lineage (Weintraub, 1993). Some of these MRF expressing somite cells will differentiate in situ and form the deep muscles of the back; however, most skeletal muscle arise from myoblasts that have migrated away from the somites to populate, for instance, the abdomen and the limb. The migrating myoblasts do not express any MRF; however, they do express another transcription factor important for myogenesis called Pax3 (Buckingham and Relaix, 2007). *Pax3*, and the related gene *Pax7*, is expressed in migrating myoblasts but it is not until these cells populate the limb that they begin to express *myf5* and *myoD*.

FGF signalling has been found to be an important regulator of myogenesis in the limb. The myogenic cells in the limb that express *myoD* and *myf5* are also known to express *FGF-R4* (Marcelle et al., 1995). Electroporation of a secreted, dominant negative form of FGF-R4 into chick limb buds interferes with muscle differentiation as shown by a down-regulation of *myoD* expression and myosin heavy chain protein (Marics et al., 2002). FGF-R4 also has been implicated as a myogenic regulator in somites. A screen for genes regulated by Pax3 in mouse embryos identified *FGF-R4* as one transcriptional target that is dependent on Pax3 for its expression in the somites. Furthermore, this work found that the negative regulator of FGF signalling, *Sprouty1*, is also directly regulated by Pax3 (Lagha et al., 2008). Inhibition of FGF signalling by overexpressing *Sprouty* in limb muscles resulted in an increase in progenitors as compared to *MRF* expressing muscle cells, suggesting a role for FGF in regulating the balance between proliferation and differentiation. However, unlike the findings in tissue culture where FGF (for the most part) promotes the proliferation of myogenic precursors such that FGF must be inhibited to allow skeletal muscle differentiation, the findings of these in vivo studies suggest that FGF activity promotes myogenic differentiation.

In zebrafish, *FGF8* is expressed in the anterior part of the early somites, while *myoD* is expressed in the posterior somites as well as in the medial, adaxial cells that line up adjacent to the midline structures (the neural tube and notochord). The adaxial cells are dependent on hedgehog signalling and will go on to form slow muscle fibres; some of these cells remain medial and express engrailed, while some will migrate from the midline to the most lateral part of the somite before differentiating. The rest of the cells in the lateral somite will form fast muscle fibres and this is dependent on FGF8 signalling (Groves et al., 2005). *acerebellar* (*ace*) mutant zebrafish, which carry a mutation in FGF8 or fish injected with an AMO directed against FGF8, or treated with the FGF inhibitor SU5402, all have reduced levels of *myoD* expression in the lateral somite. Interestingly, the levels of *Pax3/7* expression were found to be higher in zebrafish lacking FGF8 signalling, suggesting that in the absence of FGF8 and myoD, these cells stay in a progenitor state (Hammond et al., 2007).

The analysis of growth factors that activate myogenesis in *Xenopus* was accelerated by mesoderm induction studies (see section on Mesoderm Induction); indeed, the first molecular marker used in these assays was *skeletal muscle actin* (Gurdon et al., 1989). In the amphibian embryo, mesoderm is induced at the equator of the blastula embryo in response to nodal signals emanating from the vegetal hemisphere. Subsequent to induction, mesodermal cells move inside the embryo during gastrulation and eventually give rise to a wide spectrum of tissue types including skeletal muscle. Several FGF ligands are expressed very early in the nascent mesoderm and in the somites (Isaacs et al., 1995; Isaacs et al., 1992; Lea et al., 2009).

Nodal activity nodal is mimicked by the related *TGF-β* family protein activin. The expression of *FGF4* (*eFGF*) is an immediate early response to activin (Fisher, 2002) and is essential for the activation of *myoD* expression in the mesoderm (Figure 13). Moreover, it has been demonstrated that a functional FGF signalling pathway is required for activin to induce mesoderm (Cornell and Kimelman, 1994). Explants expressing a dominant negative FGFR cannot form mesoderm in response to activin, despite there being no defect in the activin signalling pathway. Activin can induce *myoD* expression in animal caps, but this requires protein synthesis (Hopwood et al., 1989); very

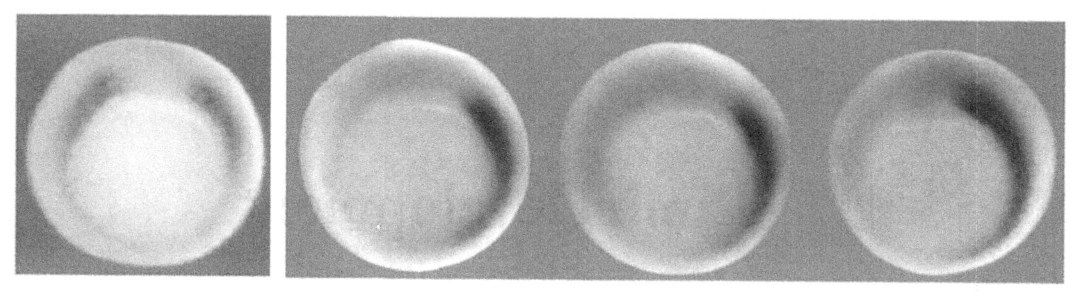

FIGURE 13: Early mesodermal expression of MyoD requires FGF4. The embryo on the left is a control *Xenopus* gastrula (NF stage 10.5) showing the expression of MyoD in the mesoderm around the blastopore. The embryos on the right have been injected into one cell at the two-cell stage with an AMO that blocks the translation of FGF4 so that only the left side of the embryo is lacking FGF4. The injected side shows that MyoD requires FGF4 for its expression in the early mesoderm, while the uninjected side provides an internal control showing normal MyoD expression. (See Fisher et al., 2002.)

likely this reflects the requirement for the production of FGF ligands to activate many mesodermal genes. Explants with impaired FGF signalling do not activate *myoD* expression in response to activin signals (Cornell and Kimelman, 1994). However, FGF4 induces *myoD* expression in animal caps in the absence of protein synthesis and is required for *myoD* expression in vivo (Fisher, 2002). These data put FGF4 at the head of the myogenic programme in *Xenopus*. There is some evidence that the molecular mechanism underlying the induction of *myoD* expression by FGF could be through MAPK-dependent inhibition of a transcriptional repressor such as Hes6 or Groucho (Burks et al., 2009; Murai et al., 2007).

In *Xenopus*, myoD mRNA and protein are present at high levels in the myogenic lineage before differentiation (Hopwood et al., 1989; Hopwood et al., 1992). In addition, a single myogenic precursor cell taken from a late gastrula embryo behaves as a determined myoblast: when transplanted to a ventral region, it differentiates as skeletal muscle (Kato and Gurdon, 1993). Therefore, these cells in the early mesoderm of the *Xenopus* gastrula represent a population skeletal muscle precursor cells. In amniotes, a similar pool of muscle progenitors is present in the medial edge of the dermomyotome as determined by lineage labeling (Denetclaw et al., 1997; Pownall et al., 2002). To maintain the stable expression of *myoD* and *myf5*, these precursor cells need to remain in contact with each other; this interaction is known as a "community effect" and is thought to help groups of muscle precursor cells coordinate their differentiation (Gurdon, 1988). When mesoderm cells are explanted from *Xenopus* gastrulae and cultured as single cells, they extinguish the expression of *myoD* and *myf5*, while intact explants or reaggregated cells express normal levels of the MRF genes. Adding FGF4 to the dispersed cell cultures rescues the expression of *myoD* and *myf5* (Standley et al., 2001); this result, together with the coexpression of MRF genes and several FGF ligands, suggests that the community effect is mediated by FGF signalling.

Limb Development

Two key signalling centres drive vertebrate limb development: the apical ectodermal ridge (AER) and the zone of polarising activity (ZPA). Embryological manipulations of the chick limb bud defined the AER as essential for the proximal-to-distal (shoulder-to-digits) outgrowth of the limb and the ZPA as the source of a morphogen that patterns anterior-to-posterior (thumb-to-pinky) axis of the limb. It may seem intuitive that the development of these two axes should be linked, as this would allow for the production of reproducible pattern: a perfect limb every time. Evidence supports this notion, and it has been shown that the AER depends on signals from the ZPA, and the ZPA depends on signals from the AER (Figure 14).

APICAL ECTODERMAL RIDGE

At about 3 days of development, limb buds form on the flank of a chick embryo. The limb bud consists of loosely associated mesoderm inside an ectodermal jacket. At the distal edge of the limb bud, a thickened ectoderm forms at the dorsal–ventral boundary (see Figure 14). This is the AER. Removal of the AER results in the cessation of limb bud out-growth and a truncated proximal–distal (PD) axis. The earlier the AER is removed, the more truncated the resulting limb is, with only the very proximal structures (humerus) forming. The later the AER is removed, the less truncated the resulting limb is, with the more distal structures forming (humerus and radius/ulna). These results can be explained by the "progress zone model." In this model, the progress zone comprises the proliferative mesodermal cells that lie just under the AER and drive the limb bud outgrowth along the PD axis. Cells leave the progress zone and differentiate in a proximal to distal direction; so, the cells of the shoulder exit the progress zone first and differentiate, later the cells that form the humerus exit the progress zone, and so on. The cells that give rise to the digits stay in the progress zone the longest time (Summerbell et al., 1973). In this way, the cells that spend the shortest time in the progress zone become proximal structures and the cells that spend the longest time in the progress zone become distal structures. Much of the data derived from chick experimental embryology fit this model very well (Tickle and Wolpert, 2002). There is, however, an alternative model for PD patterning that has been proposed more recently (Dudley et al., 2002), where the very early

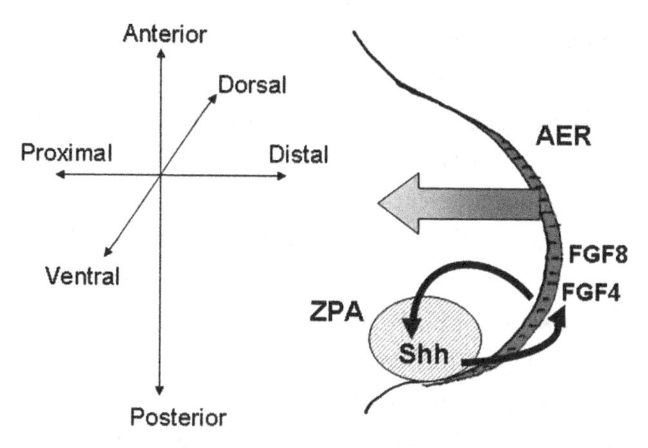

FIGURE 14: A feedback loop between the AER and the ZPA. *Shh* is expressed in the ZPA in the posterior mesoderm of the limb bud. *FGF4* is expressed in the posterior AER, while *FGF8* is expressed throughout the AER. The continued expression of *Shh* depends on FGF signalling, while the expression of *FGF4* requires Shh signals. Together, the ZPA and AER pattern the anterior–posterior and proximal–distal axes of the vertebrate limb.

limb bud is prespecified along the PD axis and the AER simply directs the expansion of these predetermined regions during limb bud out-growth.

When the AER is removed from a limb bud, further PD outgrowth stops. Both outgrowth and PD patterning can be rescued by replacing the AER with an FGF bead (Niswander et al., 1993). This indicates that it is FGF signalling that mediates the activity of the AER. Consistent with this finding, FGF8 is expressed along the entire anteroposterior length of the AER and FGF4 is expressed in the posterior half of the AER. FGF signals from the AER are thought to maintain the cells in the progress zone as an undifferentiated, proliferative mass of mesoderm that drives outgrowth of limb bud. Emphasising the absolutely central role for FGF in limb development, an excellent experiment showed that an FGF bead implanted in the interlimb flank of a chick embryo will induce the development of a complete, ectopic limb (Cohn et al., 1995). In an interesting set of follow-up experiments, these researchers implanted FGF beads into the lateral plate mesoderm in python embryos in an attempt to get limbs to grow in snakes (Cohn and Tickle, 1999)!

In mice, any essential role for each of the FGF ligands expressed in the AER has been tested. *FGF4*, *FGF8*, *FGF9*, and *FGF17* are expressed in the mouse AER. Mice lacking *FGF8* in the AER (Lewandoski et al., 2000) show a limb phenoptype where limb buds are smaller and some skeletal elements are reduced or missing; however, individually knocking out any of the other FGFs in limbs has no effect. To test the sufficiency of FGF8 for AER function, all of the FGFs were genetically removed from the limb except for FGF8. To do this, mice homozygous null for FGF9 and FGF17 were crossed with conditional knockouts where FGF4 is depleted only in the limb [mice homozygous null for FGF4 are embryonic lethal (Feldman et al., 1995)]; this creates mice with limbs lacking FGF4, FGF9, and FGF17 (Mariani et al., 2008). Despite not expressing three of the four FGF ligands that are normally present in the AER, limb development in these mice was found to be normal. This means that FGF8 is sufficient to drive and pattern PD development in the limb.

Perhaps most interestingly, this work titrated down overall levels of FGF and showed that double conditional knockouts for FGF4 and FGF8 result in very small forelimbs that, surprisingly, retain some PD pattern. When this double knockdown is combined with a mouse heterozygous FGF9, there is further reduction in forelimb development where a small humerus (stylopod) forms with a digit attached, but no radius or ulna (zeugopod) and for the most part no autopod (hand/paw). This could mean that FGFs have an instructive role in the formation of distal structures (zeugopod and autopod) during mouse limb development rather than just a permissive role in maintaining the proliferation of cells in the progress zone (Mariani et al., 2008).

However, this same study (Mariani et al., 2008) showed that there is a complete loss of hind limb development in these same FGF8/FGF4 double conditional knockout mice. This difference is due to the earlier activation of the CRE-driver in the hind limb and points to the early essential role for FGFs in limb development. In further support of this notion, another study showed that when FGFR2 was specifically depleted in the limb (using a similar method as (Mariani et al., 2008), this resulted in limbs with reduced distal outgrowth that lack digits. This phenotype is very similar to that seen in AER ablation experiments and is likely attributable to the failure of these limbs to maintain proliferation in the progress zone in the absence of AER-derived FGF signals (Yu and Ornitz, 2008).

There has been considerable debate as to what model best fits these findings: the progresss zone model where the AER-derived FGFs maintain limb bud outgrowth and positional information is determined by how long a cell remains in the PZ under the influence of FGF seems in stark contradiction with findings described above where FGFs seem to have an instructive role in dictating PD cell fate (Mariani et al., 2008). We recommend a very nice recent review that works to unify these models (Towers and Tickle, 2009).

THE ZPA REQUIRES THE AER AND VICE VERSA

The ZPA is operationally defined as a region of posterior mesoderm that, when transplanted into the anterior mesoderm of another limb, can induce mirror-image duplications of the AP axis of the limb. However, it is clear from these experiments that the resulting limb duplications are also appropriately patterned along the PD axis, indicating that the grafted tissue can direct patterning along both axes. In addition, it is known that ZPA grafts can only induce limb duplication when grafted close to the AER and that the polarising activity found in this mesoderm diminishes a few hours after and AER is extirpated. This evidence points to an important role for the ridge in promoting polarising activity, and we discuss here how FGF signalling that is central to this function.

Sonic hedgehog (Shh) is expressed in the ZPA and, like the ZPA, Shh can induce mirror-image AP axis duplication when supplied to the anterior mesoderm of another limb bud (Riddle et al. 1993). It has been shown that Shh can activate *FGF4* expression in limb ectoderm (Laufer et al., 1994). In this study it was found that ectopic Shh induced the expression of *FGF4* in the anterior AER; normally, *FGF4* expression is restricted to the posterior ridge. In the mesoderm, Shh activates the expression of *BMP2* and *HoxD11*. However, when Shh is presented to mesoderm cells that are some distance from the AER, it fails to induce the expression of these targets, pointing to a requirement for the gene products (such as FGF4) that are induced by Shh and are present in the AER. It is thought that AER-derived FGF signalling acts on the mesoderm and renders it competent to respond to Shh signalling. FGF4 is thought to play a similar role of providing competence during the induction of mesoderm in frogs (Cornell and Kimelman, 1994; Isaacs, 1997), and it is possible that this FGF function is also prevalent during limb development.

The work described above shows how Shh can activate *FGF4* expression; in addition, FGF is known to be important for the expression of *Shh*. Ectopic *Shh* expression can be activated by a combination of FGF and RA beads (Niswander et al., 1994). In limb buds where the AER has been removed, Shh expression in the ZPA is lost; however, this expression can be rescued by supplying an FGF bead. Taken together, these data point to a positive feedback loop where FGF4 maintains the expression of *Shh* and vice versa. This co-dependent relationship is reflected by the continued expression of *Shh* in the posterior mesoderm just underlying the *FGF4* expressing AER throughout the PD outgrowth of the limb. This close interdependency of the developing PD and AP axes is essential for normal limb development. However, some recent work reveals the complexity to the inner workings of this co-regulation. The ETS-box transcription factor Etv4 and Etv5 were found to be transcriptional targets of AER-derived FGFs in the limb (Mao et al., 2009; Zhang et al., 2009). This is not unexpected as ETS-box genes are often found to be activated downstream of FGF signalling. However, disruption of Etv4 or Etv5 in the limb was found not to affect PD development, but leads to polydactyly where extra posterior digits are formed, pointing to effects on the AP axis. This phenotype is found in embryos where *Shh* expression is up-regulated due to mutations in genes required to restrict *Shh* expression. It was found that disruption of Etv4/5 resulted in excessive *Shh* expression. This is very surprising because, as discussed above, FGF is a known positive regulator of Shh expression in the limb. The interpretation of these data is that FGF acts through Etv4 and Etv5 to inhibit Shh in the anterior limb mesoderm and thereby restrict polarising activity to the posterior (Mao et al., 2009; Zhang et al., 2009).

FGF and Left-Right Asymmetry

Vertebrates, including humans, typically appear symmetrical from the outside; however, internally, there is clear left–right (LR) asymmetry: the heart, stomach, and spleen are on the left, while the liver and gall bladder are on the right; the right side of the lungs has more lobes than the left, and the intestines coil counterclockwise. This patterning is set up very early during development and requires the asymmetric expression of *nodal* in the left lateral plate mesoderm. This is a conserved aspect of normal LR patterning in all vertebrates studied; however, distinct mechanisms have been found to regulate left-sided *nodal* expression in different species (reviewed in Raya and Belmonte, 2006). In addition, many vertebrates have a ciliated laterality organ. In mouse and chick, it is the node, in zebrafish it is Kupffer's vesicle (KV), and in frog it is the gastrocoel roof plate (GRP). The cilia in these laterality organs are motile and create a directional fluid flow. This directional flow is essential for LR patterning, and this has been dramatically demonstrated by exogenously manipulating the direction of the flow to impact LR asymmetry in cultured mouse embryos (Nonaka et al., 2002). Somewhat confusingly, this leftward flow of extraembryonic fluid is referred to as "nodal flow"; and although the TGFβ molecule nodal plays a conserved and essential role in LR asymmetry, "nodal flow" does not refer the protein nodal but rather to the node.

FGF8 has been implicated in the early processes that regulate LR asymmetry (Figure 15). Mice that are null for FGF8 do not develop through gastrula stages; however, in mice heterozygous for a hypomorphic allele ($FGF8^{neo/-}$), about half were found to be abnormal in their LR patterning and lack *nodal* expression in the left lateral plate mesoderm (Meyers and Martin, 1999). In amniotes, FGF8 is expressed in the primitive streak, and in chick FGF8 expression extends anteriorly from the streak on the right-hand side of Hensen's node during very early somite stages where it inhibits the expression of *nodal* (Boettger et al., 1999). Implanting FGF beads into the left lateral plate mesoderm in chick represses *nodal* expression and randomises the direction of heart looping. This is not seen in the mouse; indeed, the finding that FGF8 mutant mice do not express *nodal* and that implanting an FGF8 bead will induce *nodal* expression in the right lateral plate mesoderm is in striking contrast to the results in chick (Meyers and Martin, 1999). These contradictory data were probed further by investigating effects of FGF8 on LR asymmetry in cultured rabbit embryos, which, like chick, develop as a flat disc, while mice develop as a cylinder. Similar to mice, FGF8 was

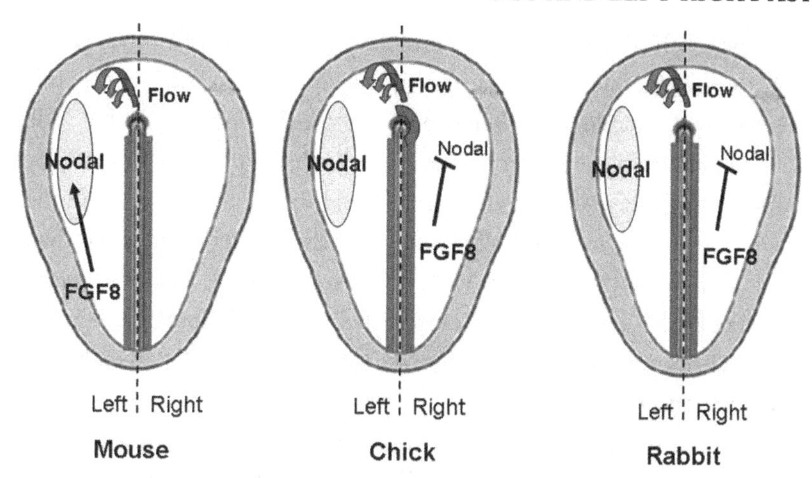

FIGURE 15: FGF8 regulates *Nodal* expression differently in chick and mouse embryos during LR axis specification. In the mouse, FGF8 (shown in red) is expressed symmetrically in the primitive streak and is necessary and sufficient for the Expression of *Nodal* in the left lateral plate mesoderm (shown in blue stripes). In the chick, FGF8 expression extends anteriorly on the right and represses FGF8 expression in the right lateral plate mesoderm. Interestingly, the role of FGF8 in the rabbit is similar to its role in chick, while its expression is symmetrical, like in the mouse.

expressed symmetrically in the primitive streak in rabbits; however, similar to chick, FGF8 beads repressed endogenous *nodal* expression in the left lateral plate mesoderm. Application of SU5402 to the right lateral plate mesoderm resulted in bilateral nodal expression, indicating that in rabbit, as in chick, FGF8 acts to repress nodal expression (Fischer et al., 2002). The surprisingly different roles for FGF8 in two mammals point to the importance of the relative anatomy of the embryo in determining FGF8 function. Interestingly, the ability of FGF8 to repress nodal on the right in rabbits was found to require intercellular communication through gap junctions; the leftward nodal flow is thought to relieve FGF repression on the left by locally restricting gap junction communication (Feistel and Blum, 2008).

Cilia-driven nodal flow is considered the earliest symmetry-breaking event during LR patterning because mice lacking motile cilia (Nonaka et al., 1998) have disrupted expression of LR markers and impaired directionality of heart looping. By labelling lipid membranes, Tanaka et al. (2005) were able to visualise remarkable particles less than 0.5 μm in diameter that carry Sonic hedgehog protein and RA that were being transported toward the left in the ventral node of one- to three-somite-stage mouse embryos. The release of these nodal vesicular particles (NVPs) from microvilli was found to be dependent on FGF signalling; as in embryos treated with SU5402, there is a failure to launch NVPs and they remain associated with the microvilli membrane in an intermediate form (Tanaka et al., 2005).

As in mice, fish mutant for *fgf8* show defects in organ and pharyngeal skeleton laterality. In zebrafish, the ciliated laterality organ KV is absent in approximately one third of *fgf8* mutants (*acerebellar*) pointing to a role for fgf8 in supporting the development of KV (Albertson and Yelick, 2005). KV is derived from the dorsal forerunner cells that express *notail* (*ntl*), a gene known to dependent on FGF signalling (Griffin et al., 1995), and the expression of *ntl* in the dorsal forerunner cells is deficient in *ace* mutant fish (Albertson and Yelick, 2005) suggesting that FGF8 is required in KV progenitors. In another study, AMOs were used to knock down fgf8 (as well as some downstream effectors of FGF signalling), which revealed a dramatic loss of cilia in KV and random expression of laterality markers including *southpaw* (*spw*) and *lefty* (*lfty*) in embryos deficient in FGF8 signalling (Hong and Dawid, 2009).

LR patterning was also investigated at a later time point, during the development of brain asymmetry in the diencephalic epithalamus (Regan et al., 2009). In zebrafish, parapineal precursors move from their bilateral positions at the midline and migrate leftward; this event leads to the left-specific organisation of this side of the brain. This migration is dependent on the left-sided nodal expression in the lateral plate mesoderm and, in the absence of *nodal*, there is equal likelihood of the parapineal precursors moving to the left or to the right (Concha et al., 2000). This indicates that there is still brain asymmetry without *nodal*, although it can have an opposite orientation. *Ace(fgf8)* mutant fish were found to have symmetrical epithalamus, and their pineal precursors do not migrate from the midline (Regan et al., 2009) indicating that fgf8 is important for brain asymmetry. Implanting an FGF bead into *ace* mutants allows some of the parapineal cells to migrate, and most of these migrate to the left, probably because there is still left expression of nodal in the epithalmus of the majority of *ace* embryos. The model proposed by the authors suggests that the midline expression of *fgf8* makes the parapineal cells unstable so that they move. If there is no *nodal*, they lack direction and migrate either to the left or to the right; however, if there is no *fgf8*, then the cells do not migrate at all. This suggests that during the establishment of asymmetric patterning, simply generating asymmetry is a first step and determining the directionality of asymmetry is a subsequent event.

In zebrafish, FGFRs 1, 2, and 3 have been found on the cilia in KV using specific antibodies (Tanaka et al., 2005) and transcripts for FGFR-1 are localised to KV, the presomitic and lateral plate mesoderm, and distinct regions of the brain at the six-somite stage (Neugebauer et al., 2009). AMOs mediate knockdown of FGFR1 results in the bilateral expression of *southpaw*, which is normally expressed only in the left lateral plate mesoderm. Specifically depleting FGFR-1 in KV results in the same effects on *spaw* expression, and while there was no effect on the number or polarity of the cilia in KV, the cilia were significantly shorter in FGFR-1 knockdown zebrafish. The findings that cilia length was shortened by the disruption of FGF signalling in different developing organs that require FGF for their morphogenesis (such as the otic vesicle and the pronephros) and in laterality organs in another species (the *Xenopus* GRP) suggest that maintaining cilia length is an important mechanism through which FGF regulates developmental processes. In addition, cilia are known to play a role in modulating cell signalling through many pathways that include Wnt and Shh, and these pathways are known to interact with FGF signalling. This could mean that effects of FGF on cilia length may be another mechanism by which FGF signalling contributes to the regulation of other signal transduction pathways.

Perspectives

The many roles that FGF signalling plays during vertebrate development illustrated by the examples in this review reveal the progress that has been made over the last 25 years in our understanding of the biological functions of this pathway. In the late 1980s, the discovery that purified FGF2, a protein known to be present in the adult, could mimic a significant signalling event in the embryo, mesoderm induction, had a profound effect on how the regulation of development was viewed and set the scene for much subsequent research. As is hopefully apparent from this review, the FGF signalling pathway represents an ancient regulatory cassette that has been deployed multiple times during animal evolution to regulate diverse processes. In some contexts, multiple FGF ligands and receptors are used in the development of the same organ system. In common with other families of signalling molecules, it is clear that FGFs have overlapping activities, which, given the shared pathway components, is perhaps not surprising. However, many questions remain concerning the generation of specific biological output from the plethora of ligand and receptor combinations that the complex FGF system allows. Much is still to be learnt concerning the signalling and transcriptional events downstream of the FGFRs. Furthermore, in recent years, it has become apparent that the FGF pathway is subject to modulation at multiple levels, from the extracellular environment to the nucleus. Understanding the integration of these modulatory systems will be key to a full understanding of how FGF signalling operates in combination with other signalling pathways to regulate the exquisitely refined yet robust process of animal development.

References

Abler, L. L., Mansour, S. L., Sun, X., 2009. Conditional gene inactivation reveals roles for Fgf10 and Fgfr2 in establishing a normal pattern of epithelial branching in the mouse lung, *Dev Dyn*, 238, pp. 1999–2013.

Abu-Issa, R., Smyth, G., Smoak, I., Yamamura, K., Meyers, E. N., 2002. Fgf8 is required for pharyngeal arch and cardiovascular development in the mouse, *Development*, 129, pp. 4613–25.

Acampora, D., Avantaggiato, V., Tuorto, F., Briata, P., Corte, G., Simeone, A., 1998. Visceral endoderm-restricted translation of Otx1 mediates recovery of Otx2 requirements for specification of anterior neural plate and normal gastrulation, *Development*, 125, pp. 5091–5104.

Affolter, M., Zeller, R., Caussinus, E., 2009. Tissue remodelling through branching morphogenesis, *Nat Rev Mol Cell Biol*, 10, pp. 831–42.

Agius, E., Oelgeschlager, M., Wessely, O., Kemp, C., De Robertis, E. M., 2000. Endodermal Nodal-related signals and mesoderm induction in *Xenopus*, *Development*, 127, pp. 1173–83.

Albertson, R. C., Yelick, P. C., 2005. Roles for fgf8 signaling in left–right patterning of the visceral organs and craniofacial skeleton, *Dev Biol*, 283, pp. 310–21.

Alonso, A., Sasin, J., Bottini, N., Friedberg, I., Osterman, A., Godzik, A., Hunter, T., Dixon, J., Mustelin, T., 2004. Protein tyrosine phosphatases in the human genome, *Cell*, 117, pp. 699–711.

Amaya, E., Musci, T. J., Kirschner, M. W., 1991. Expression of a dominant negative mutant of the FGF receptor disrupts mesoderm formation in *Xenopus* embryos, *Cell*, 66, pp. 257–70.

Amaya, E., Stein, P. A., Musci, T. J., Kirschner, M. W., 1993. FGF signalling in the early specification of mesoderm in *Xenopus*, *Development*, 118, pp. 477–87.

Arman, E., Haffner-Krausz, R., Gorivodsky, M., Lonai, P., 1999. Fgfr2 is required for limb outgrowth and lung-branching morphogenesis, *Proc Natl Acad Sci USA*, 96, pp. 11895–99.

Basson, M. A., Echevarria, D., Ahn, C. P., Sudarov, A., Joyner, A. L., Mason, I. J., Martinez, S., Martin, G. R., 2008. Specific regions within the embryonic midbrain and cerebellum require different levels of FGF signaling during development, *Development*, 135, pp. 889–98.

Beenken, A., Mohammadi, M., 2009. The FGF family: biology, pathophysiology and therapy, *Nat Rev Drug Discov*, 8, pp. 235–53.

Bertrand, S., Somorjai, I., Garcia-Fernandez, J., Lamonerie, T., Escriva, H., 2009. FGFRL1 is a neglected putative actor of the FGF signalling pathway present in all major metazoan phyla, *BMC Evol Biol*, 9, p. 226.

Bertrand, V., Hudson, C., Caillol, D., Popovici, C., Lemaire, P., 2003. Neural tissue in ascidian embryos is induced by FGF9/16/20, acting via a combination of maternal GATA and Ets transcription factors, *Cell*, 115, pp. 615–27.

Bink, R. J., Habuchi, H., Lele, Z., Dolk, E., Joore, J., Rauch, G. J., Geisler, R., Wilson, S. W., den Hertog, J., Kimata, K., Zivkovic, D., 2003. Heparan sulfate 6-*o*-sulfotransferase is essential for muscle development in zebrafish, *BMC Evol Biol*, 2009; 9, p. 226.

Birsoy, B., Kofron, M., Schaible, K., Wylie, C., Heasman, J., 2006. Vg 1 is an essential signaling molecule in *Xenopus* development, *Development*, 133, pp. 15–20.

Boettger, T., Wittler, L., Kessel, M., 1999. FGF8 functions in the specification of the right body side of the chick, *Current Biology*, 9, pp. 277–80.

Bottcher, R. T., Niehrs, C., 2005. Fibroblast growth factor signaling during early vertebrate development, *Endocr Rev*, 26, pp. 63–77.

Bottcher, R. T., Pollet, N., Delius, H., Niehrs, C., 2004. The transmembrane protein XFLRT3 forms a complex with FGF receptors and promotes FGF signalling, *Nat Cell Biol*, 6, pp. 38–44.

Branney, P. A., Faas, L., Steane, S. E., Pownall, M. E., Isaacs, H. V., 2009. Characterisation of the fibroblast growth factor dependent transcriptome in early development, *PLoS ONE*, 4, p. e4951.

Buckingham, M., Meilhac, S., Zaffran, S., 2005. Building the mammalian heart from two sources of myocardial cells, *Nat Rev Genet*, 6, pp. 826–35.

Buckingham, M., Relaix, F., 2007. The role of Pax genes in the development of tissues and organs: Pax3 and Pax7 regulate muscle progenitor cell functions, *Annu Rev Cell Dev Biol*, 23, pp. 645–73.

Burks, P. J., Isaacs, H. V., Pownall, M. E., 2009. FGF signalling modulates transcriptional repression by *Xenopus* groucho-related-4, *Biol Cell*, 101, pp. 301–08.

Cabrita, M. A., Christofori, G., 2008. Sprouty proteins, masterminds of receptor tyrosine kinase signaling, *Angiogenesis*, 11, pp. 53–62.

Cadwallader, A. B., Yost, H. J., 2006. Combinatorial expression patterns of heparan sulfate sulfotransferases in zebrafish: II. The 6-*O*-sulfotransferase family, *Dev Dyn*, 235, pp. 3432–37.

Casci, T., Vinos, J., Freeman, M., 1999. Sprouty, an intracellular inhibitor of Ras signaling, *Cell*, 96, pp. 655–65.

Chen, E., Stringer, S. E., Rusch, M. A., Selleck, S. B., Ekker, S. C., 2005. A unique role for 6-O sulfation modification in zebrafish vascular development, *Dev Biol*, 284, pp. 364–76.

Chen, Y., Mohammadi, M., Flanagan, J. G., 2009. Graded levels of FGF protein span the midbrain and can instruct graded induction and repression of neural mapping labels, *Neuron*, 62, pp. 773–80.

Christen, B., Slack, J. M., 1997. FGF-8 is associated with anteroposterior patterning and limb regeneration in *Xenopus*, *Dev Biol*, 192, pp. 455–66.

Christen, B., Slack, J. M., 1999. Spatial response to fibroblast growth factor signalling in *Xenopus* embryos, *Development*, 126, pp. 119–25.

Chung, H. A., Hyodo-Miura, J., Kitayama, A., Terasaka, C., Nagamune, T., Ueno, N., 2004. Screening of FGF target genes in *Xenopus* by microarray: temporal dissection of the signalling pathway using a chemical inhibitor, *Genes Cells*, 9, pp. 749–61.

Chung, H. A., Hyodo-Miura, J., Nagamune, T., Ueno, N., 2005. FGF signal regulates gastrulation cell movements and morphology through its target NRH, *Dev Biol*, 282, pp. 95–110.

Cinnamon, E., Helman, A., Ben-Haroush Schyr, R., Orian, A., Jimenez, G., Paroush, Z., 2008. Multiple RTK pathways downregulate Groucho-mediated repression in *Drosophila* embryogenesis, *Development*, 135, pp. 829–37.

Ciruna, B., Rossant, J., 2001. FGF signaling regulates mesoderm cell fate specification and morphogenetic movement at the primitive streak, *Dev Cell*, 1, pp. 37–49.

Ciruna, B. G., Schwartz, L., Harpal, K., Yamaguchi, T. P., Rossant, J., 1997. Chimeric analysis of fibroblast growth factor receptor-1 (Fgfr1) function: a role for FGFR1 in morphogenetic movement through the primitive streak, *Development*, 124, pp. 2829–41.

Clegg, C. H., Linkhart, T. A., Olwin, B. B., Hauschka, S. D., 1987. Growth factor control of skeletal muscle differentiation: commitment to terminal differentiation occurs in G1 phase and is repressed by fibroblast growth factor, *J Cell Biol*, 105, pp. 949–56.

Cohn, M. J., Izpisua-Belmonte, J. C., Abud, H., Heath, J. K., Tickle, C., 1995. Fibroblast growth factors induce additional limb development from the flank of chick embryos, *Cell*, 80, pp. 739–46.

Cohn, M. J., Tickle, C., 1999. Developmental basis of limblessness and axial patterning in snakes, *Nature*, 399, pp. 474–79.

Colvin, J. S., Bohne, B. A., Harding, G. W., McEwen, D. G., Ornitz, D. M., 1996. Skeletal overgrowth and deafness in mice lacking fibroblast growth factor receptor 3, *Nat Genet*, 12, pp. 390–97.

Concha, M. L., Burdine, R. D., Russell, C., Schier, A. F., Wilson, S. W., 2000. A nodal signaling pathway regulates the laterality of neuroanatomical asymmetries in the zebrafish forebrain, *Neuron*, 28, pp. 399–409.

Cornell, R. A., Kimelman, D., 1994. Activin-mediated mesoderm induction requires FGF, *Development*, 120, pp. 453–462.

Crossley, P. H., Martinez, S., Martin, G. R., 1996. Midbrain development induced by FGF8 in the chick embryo, *Nature*, 380, pp. 66–68.

Dailey, L., Ambrosetti, D., Mansukhani, A., Basilico, C., 2005. Mechanisms underlying differential responses to FGF signaling, *Cytokine Growth Factor Rev*, 16, pp. 233–47.

Daudet, N., Lewis, J., 2005. Two contrasting roles for Notch activity in chick inner ear development: specification of prosensory patches and lateral inhibition of hair-cell differentiation, *Development*, 132, pp. 541–51.

Delaune, E., Lemaire, P., Kodjabachian, L., 2005. Neural induction in *Xenopus* requires early FGF signalling in addition to BMP inhibition, *Development*, 132, pp. 299–310.

Delfini, M. C., Dubrulle, J., Malapert, P., Chal, J., Pourquie, O., 2005. Control of the segmentation process by graded MAPK/ERK activation in the chick embryo, *Proc Natl Acad Sci USA*, 102, pp. 11343–48.

Denetclaw, W. F., Jr., Christ, B., Ordahl, C. P., 1997. Location and growth of epaxial myotome precursor cells, *Development*, 124, pp. 1601–10.

Dequeant, M. L., Glynn, E., Gaudenz, K., Wahl, M., Chen, J., Mushegian, A., Pourquie, O., 2006. A complex oscillating network of signaling genes underlies the mouse segmentation clock, *Science*, 314, pp. 1595–98.

Dequeant, M. L., Pourquie, O., 2008. Segmental patterning of the vertebrate embryonic axis, *Nat Rev Genet*, 9, pp. 370–82.

Dhoot, G. K., Gustafsson, M. K., Ai, X., Sun, W., Standiford, D. M., Emerson, C. P., Jr., 2001. Regulation of Wnt signaling and embryo patterning by an extracellular sulfatase, *Science*, 293, pp. 1663–66.

Dickinson, R. J., Eblaghie, M. C., Keyse, S. M., Morriss-Kay, G. M., 2002. Expression of the ERK-specific MAP kinase phosphatase PYST1/MKP3 in mouse embryos during morphogenesis and early organogenesis, *Mech Dev*, 113, pp. 193–96.

Diez del Corral, R., Breitkreuz, D. N., Storey, K. G., 2002. Onset of neuronal differentiation is regulated by paraxial mesoderm and requires attenuation of FGF signalling, *Development*, 129, pp. 1681–91.

Diez del Corral, R., Olivera-Martinez, I., Goriely, A., Gale, E., Maden, M., Storey, K., 2003. Opposing FGF and retinoid pathways control ventral neural pattern, neuronal differentiation, and segmentation during body axis extension, *Neuron*, 40, pp. 65–79.

Diez del Corral, R., Storey, K. G., 2004. Opposing FGF and retinoid pathways: a signalling switch that controls differentiation and patterning onset in the extending vertebrate body axis, *Bioessays*, 26, pp. 857–69.

Dubrulle, J., McGrew, M. J., Pourquie, O., 2001. FGF signaling controls somite boundary position and regulates segmentation clock control of spatiotemporal Hox gene activation, *Cell*, 106, pp. 219–32.

Dudley, A. T., Ros, M. A., Tabin, C. J., 2002. A re-examination of proximodistal patterning during vertebrate limb development, *Nature*, 418, pp. 539–44.

Eblaghie, M. C., Lunn, J. S., Dickinson, R. J., Munsterberg, A. E., Sanz-Ezquerro, J. J., Farrell, E. R., Mathers, J., Keyse, S. M., Storey, K., Tickle, C., 2003. Negative feedback regulation of FGF signaling levels by Pyst1/MKP3 in chick embryos, *Curr Biol*, 13, pp. 1009–18.

Esch, F., Baird, A., Ling, N., Ueno, N., Hill, F., Denoroy, L., Klepper, R., Gospodarowicz, D., Bohlen, P., Guillemin, R., 1985. Primary structure of bovine pituitary basic fibroblast growth factor (FGF) and comparison with the amino-terminal sequence of bovine brain acidic FGF, *Proc Natl Acad Sci USA*, 82, pp. 6507–11.

Faas, L., Isaacs, H. V., 2009. Overlapping functions of Cdx1, Cdx2, and Cdx4 in the development of the amphibian *Xenopus tropicalis*, *Dev Dyn*, 238, pp. 835–52.

Feistel, K., Blum, M., 2008. Gap junctions relay FGF8-mediated right-sided repression of Nodal in rabbit, *Dev Dyn*, 237, pp. 3516–27.

Feldman, B., Poueymirou, W., Papaioannou, V. E., DeChiara, T. M., Goldfarb, M., 1995. Requirement of FGF-4 for postimplantation mouse development, *Science*, 267, pp. 246–49.

Fischer, A., Viebahn, C., Blum, M., 2002. FGF8 acts as a right determinant during establishment of the left–right axis in the rabbit, *Curr Biol*, 12, pp. 1807–16.

Fisher, M. E., Isaacs, H.V. and Pownall, M.E., 2002. eFGF is required for the activation of XmyoD in the myogenic cell lineage of *Xenopus laevis*, *Development*, 129, pp. 1307–15.

Fletcher, R. B., Baker, J. C., Harland, R. M., 2006. FGF8 spliceforms mediate early mesoderm and posterior neural tissue formation in *Xenopus*, *Development*, 133, pp. 1703–14.

Fletcher, R. B., Harland, R. M., 2008. The role of FGF signaling in the establishment and maintenance of mesodermal gene expression in *Xenopus*, *Dev Dyn*, 237, pp. 1243–54.

Freeman, S. D., Moore, W. M., Guiral, E. C., Holme, A., Turnbull, J. E., Pownall, M. E., 2008. Extracellular regulation of developmental cell signaling by XtSulf1, *Dev Biol*, 320, pp. 436–445.

Freter, S., Muta, Y., Mak, S. S., Rinkwitz, S., Ladher, R. K., 2008. Progressive restriction of otic fate: the role of FGF and Wnt in resolving inner ear potential, *Development*, 135, pp. 3415–24.

Furthauer, M., Lin, W., Ang, S. L., Thisse, B., Thisse, C., 2002. Sef is a feedback-induced antagonist of Ras/MAPK-mediated FGF signalling, *Nat Cell Biol*, 4, pp. 170–74.

Furthauer, M., Van Celst, J., Thisse, C., Thisse, B., 2004. Fgf signalling controls the dorsoventral patterning of the zebrafish embryo, *Development*, 131, pp. 2853–64.

Gerber, S. D., Steinberg, F., Beyeler, M., Villiger, P. M., Trueb, B., 2009. The murine Fgfrl1 receptor is essential for the development of the metanephric kidney, *Dev Biol*, 335, pp. 106–19.

Godsave, S. F., Slack, J. M. W., 1989. Clonal analysis of mesoderm induction in *Xenopus laevis*, *Dev Biol*, 134, pp. 486–90.

Gomez, A. R., Lopez-Varea, A., Molnar, C., de la Calle-Mustienes, E., Ruiz-Gomez, M., Gomez-Skarmeta, J. L., de Celis, J. F., 2005. Conserved cross-interactions in *Drosophila* and *Xenopus* between Ras/MAPK signaling and the dual-specificity phosphatase MKP3, *Dev Dyn*, 232, pp. 695–708.

Gospodarowicz, D., 1975. Purification of a fibroblast growth factor from bovine pituitary, *J Biol Chem*, 250, pp. 2515–20.

Gospodarowicz, D., Ill, C. R., Birdwell, C. R., 1977. Effects of fibroblast and epidermal growth factors on ovarian cell proliferation in vitro. I. Characterization of the response of granulosa cells to FGF and EGF, *Endocrinology*, 100, pp. 1108–20.

Gospodarowicz, D., Moran, J. S., 1975. Mitogenic effect of fibroblast growth factor on early passage cultures of human and murine fibroblasts, *J Cell Biol*, 66, pp. 451–57.

Griffin, K., Patient, R., Holder, N., 1995. Analysis of FGF function in normal and no tail zebrafish embryos reveals separate mechanisms for formation of the trunk and the tail, *Development*, 121, pp. 2983–94.

Groves, J. A., Hammond, C. L., Hughes, S. M., 2005. Fgf8 drives myogenic progression of a novel lateral fast muscle fibre population in zebrafish, *Development*, 132, 4211–22.

Guo, Q., Li, K., Sunmonu, N. A., Li, J. Y., 2010. Fgf8b-containing spliceforms, but not Fgf8a, are essential for Fgf8 function during development of the midbrain and cerebellum, *Dev Biol*, 338. pp. 183–92.

Gurdon, J. B., 1988. A community effect in animal development, *Nature*, 336, pp. 772–74.

Gurdon, J. B., Mohun, T. J., Sharpe, C. R., Taylor, M. V., 1989. Embryonic induction and muscle gene activation, *Trends Genet*, 5, pp. 51–56.

Hacohen, N., Kramer, S., Sutherland, D., Hiromi, Y., Krasnow, M. A., 1998. sprouty encodes a novel antagonist of FGF signaling that patterns apical branching of the *Drosophila* airways, *Cell*, 92, pp. 253–63.

Hall, H., Walsh, F. S., Doherty, P., 1996. Review: a role for the FGF receptor in the axonal growth response stimulated by cell adhesion molecules? *Cell Adhes Commun*, 3, pp. 441–50.

Hammond, C. L., Hinits, Y., Osborn, D. P., Minchin, J. E., Tettamanti, G., Hughes, S. M., 2007. Signals and myogenic regulatory factors restrict pax3 and pax7 expression to dermomyotome-like tissue in zebrafish, *Dev Biol*, 302, pp. 504–21.

Hanafusa, H., Torii, S., Yasunaga, T., Nishida, E., 2002. Sprouty1 and Sprouty2 provide a control mechanism for the Ras/MAPK signalling pathway, *Nat Cell Biol*, 4, pp. 850–58.

Hardcastle, Z., Chalmers, A. D., Papalopulu, N., 2000. FGF-8 stimulates neuronal differentiation through FGFR-4a and interferes with mesoderm induction in *Xenopus* embryos, *Curr Biol*, 10, pp. 1511–14.

Hardcastle, Z., Papalopulu, N., 2000. Distinct effects of XBF-1 in regulating the cell cycle inhibitor p27(XIC1) and imparting a neural fate, *Development*, 127, pp. 1303–14.

Haremaki, T., Tanaka, Y., Hongo, I., Yuge, M., Okamoto, H., 2003. Integration of multiple signal transducing pathways on Fgf response elements of the *Xenopus* caudal homologue Xcad3, *Development*, 130, pp. 4907–17.

Hashimoto, M., Sagara, Y., Langford, D., Everall, I. P., Mallory, M., Everson, A., Digicaylioglu, M., Masliah, E., 2002. Fibroblast growth factor 1 regulates signaling via the glycogen synthase kinase-3beta pathway. Implications for neuroprotection, *BMC Evol Biol*, 9, p. 226.

Hasson, P., Paroush, Z., 2006. Crosstalk between the EGFR and other signalling pathways at the level of the global transcriptional corepressor Groucho/TLE, *Br J Cancer*, 94, pp. 771–75.

Hebert, J. M., Fishell, G., 2008. The genetics of early telencephalon patterning: some assembly required, *Nat Rev Neurosci*, 9, pp. 678–85.

Hemmati-Brivanlou, A., Melton, D., 1997. Vertebrate neural induction, *Annu Rev Neurosci*, 20, pp. 43–60.

Hild, M., Dick, A., Rauch, G. J., Meier, A., Bouwmeester, T., Haffter, P., Hammerschmidt, M., 1999. The smad5 mutation somitabun blocks Bmp2b signaling during early dorsoventral patterning of the zebrafish embryo, *Development*, 126, pp. 2149–59.

Hong, S. K., Dawid, I. B., 2009. FGF-dependent left–right asymmetry patterning in zebrafish is mediated by Ier2 and Fibp1, *Proc Natl Acad Sci USA*, 106, pp. 2230–35.

Hongo, I., Kengaku, M., Okamoto, H., 1999. FGF signaling and the anterior neural induction in *Xenopus*, *Dev Biol*, 216, pp. 561–81.

Hopwood, N. D., Pluck, A., Gurdon, J. B., 1989. MyoD expression in the forming somites is an early response to mesoderm induction in *Xenopus* embryos, *EMBO Journal*, 8, pp. 3409–17.

Hopwood, N. D., Pluck, A., Gurdon, J. B., Dilworth, S. M., 1992. Expression of XMyoD protein in early *Xenopus laevis* embryos, *Development*, 114, pp. 31–38.

Horowitz, A., Simons, M., 2008. Branching morphogenesis, *Circ Res*, 103, pp. 784–95.

Houart, C., Westerfield, M., Wilson, S. W., 1998. A small population of anterior cells patterns the forebrain during zebrafish gastrulation, *Nature*, 391, pp. 788–92.

Isaacs, H., 1997. New perspectives on the role of the fibroblast growth factor family in amphibian development, *Cell Mol Life Sci*, 53, pp. 350–61.

Isaacs, H. V., Pownall, M. E., Slack, J. M. W., 1994. eFGF regulates Xbra expression during *Xenopus* gastrulation, *EMBO Journal*, 13, pp. 4469–81.

Isaacs, H. V., Pownall, M. E., Slack, J. M. W., 1995. eFGF is expressed in the dorsal midline of *Xenopus laevis*, *Int J Dev Biol*, 39, pp. 575–79.

Isaacs, H. V., Pownall, M. E., Slack, J. M. W., 1998. Regulation of Hox gene expression and posterior development by the *Xenopus* caudal homologue Xcad3, *EMBO J*, 17, pp. 3413–27.

Isaacs, H. V., Tannahill, D., Slack, J. M. W., 1992. Expression of a novel FGF in the *Xenopus* embryo. A new candidate inducing factor for mesoderm formation and anteroposterior specification, *Development*, 114, pp. 711–20.

Israsena, N., Hu, M., Fu, W., Kan, L., Kessler, J. A., 2004. The presence of FGF2 signaling determines whether beta-catenin exerts effects on proliferation or neuronal differentiation of neural stem cells, *Dev Biol*, 268, pp. 220–31.

Itoh, N., Ornitz, D. M., 2004. Evolution of the Fgf and Fgfr gene families, *Trends Genet*, 20, pp. 563–69.

Jeanes, A., Gottardi, C. J., Yap, A. S., 2008. Cadherins and cancer: how does cadherin dysfunction promote tumor progression? *Oncogene*, 27, pp. 6920–29.

Johnson, D. E., Lu, J., Chen, H., Werner, S., Williams, L. T., 1991. The human fibroblast growth factor receptor genes: a common structural arrangement underlies the mechanisms for generating receptor forms that differ in their third immunoglobulin domain, *Mol Cell Biol*, 11, pp. 4627–34.

Jope, R. S., Johnson, G. V., 2004. The glamour and gloom of glycogen synthase kinase-3, *Trends Biochem Sci*, 29, pp. 95–102.

Kamimura, K., Fujise, M., Villa, F., Izumi, S., Habuchi, H., Kimata, K., Nakato, H., 2001. *Drosophila* heparan sulfate 6-*O*-sulfotransferase (dHS6ST) gene. Structure, expression, and function in the formation of the tracheal system, *J Biol Chem*, 276, pp. 17014–21.

Kamimura, K., Koyama, T., Habuchi, H., Ueda, R., Masu, M., Kimata, K., Nakato, H., 2006. Specific and flexible roles of heparan sulfate modifications in *Drosophila* FGF signaling, *J Cell Biol*, 174, pp. 773–78.

Kato, K., Gurdon, J. B., 1993. Single-cell transplantation determines the time when *Xenopus* muscle precursor cells acquire a capacity for autonomous differentiation, *Proceedings of the National Academy of Sciences of the United States of America*, 90, pp. 1310–14.

Katoh, M., 2006. Cross-talk of WNT and FGF signaling pathways at GSK3beta to regulate beta-catenin and SNAIL signaling cascades, *Cancer Biol Ther*, 5, pp. 1059–64.

Keenan, I. D., Sharrard, R. M., Isaacs, H. V., 2006. FGF signal transduction and the regulation of Cdx gene expression, *Dev Biol*, 299, pp. 478–88.

Keyse, S. M., 2000. Protein phosphatases and the regulation of mitogen-activated protein kinase signalling, *Curr Opin Cell Biol*, 12, pp. 186–92.

Khokha, M. K., Yeh, J., Grammer, T. C., Harland, R. M., 2005. Depletion of three BMP antagonists from Spemann's organizer leads to a catastrophic loss of dorsal structures, *Dev Cell*, 8, pp. 401–11.

Kikuchi, A., Yamamoto, H., Sato, A., 2009. Selective activation mechanisms of Wnt signaling pathways, *Trends Cell Biol*, 19, pp. 119–29.

Kim, S. H., Yamamoto, A., Bouwmeester, T., Agius, E., Robertis, E. M., 1998. The role of paraxial protocadherin in selective adhesion and cell movements of the mesoderm during *Xenopus* gastrulation, *Development*, 125, pp. 4681–90.

Kimelman, D., Kirschner, M., 1987. Synergistic induction of mesoderm by FGF and TGF-beta and the identification of an mRNA coding for FGF in the early *Xenopus* embryo, *Cell*, 51, pp. 869–77.

Knights, V., Cook, S. J., 2010. De-regulated FGF receptors as therapeutic targets in cancer, *Pharmacol Ther*, 125, pp. 105–17.

Kofron, M., Demel, T., Xanthos, J., Lohr, J., Sun, B., Sive, H., Osada, S., Wright, C., Wylie, C., Heasman, J., 1999. Mesoderm induction in *Xenopus* is a zygotic event regulated by maternal VegT via TGFbeta growth factors, *Development*, 126, pp. 5759–70.

Konigsberg, I. R., 1963. Clonal analysis of myogenesis, *Science*, 140, pp. 1273–84.

Kouhara, H., Hadari, Y., Spivak-Kroizman, T., Schilling, J., Bar-Sagi, D., Lax, I., Schlessinger, J., 1997. A lipid-anchored Grb2-binding protein that links FGF-receptor activation to the Ras/MAPK signaling pathway, *Cell*, 89, pp. 693–702.

Krejci, P., Prochazkova, J., Bryja, V., Kozubik, A., Wilcox, W. R., 2009. Molecular pathology of the fibroblast growth factor family, *Hum Mutat*, 30, pp. 1245–55.

Kunath, T., Saba-El-Leil, M. K., Almousailleakh, M., Wray, J., Meloche, S., Smith, A., 2007. FGF stimulation of the Erk1/2 signalling cascade triggers transition of pluripotent embryonic stem cells from self-renewal to lineage commitment, *Development*, 134, pp. 2895–902.

Kuroda, H., Fuentealba, L., Ikeda, A., Reversade, B., De Robertis, E. M., 2005. Default neural induction: neuralization of dissociated *Xenopus* cells is mediated by Ras/MAPK activation, *Genes Dev*, 19, pp. 1022–27.

LaBonne, C., Whitman, M., 1994. Mesoderm induction by activin requires FGF-mediated intracellular signals, *Development*, 120, pp. 463–72.

Lacy, S. E., Bonnemann, C. G., Buzney, E. A., Kunkel, L. M., 1999. Identification of FLRT1, FLRT2, and FLRT3: a novel family of transmembrane leucine-rich repeat proteins, *Genomics*, 62, pp. 417–26.

Ladher, R. K., Wright, T. J., Moon, A. M., Mansour, S. L., Schoenwolf, G. C., 2005. FGF8 initiates inner ear induction in chick and mouse, *Genes Dev*, 19, pp. 603–13.

Lagha, M., Kormish, J. D., Rocancourt, D., Manceau, M., Epstein, J. A., Zaret, K. S., Relaix, F., Buckingham, M. E., 2008. Pax3 regulation of FGF signaling affects the progression of embryonic progenitor cells into the myogenic program, *Genes Dev*, 22, pp. 1828–37.

Lai, J., Chien, J., Staub, J., Avula, R., Greene, E. L., Matthews, T. A., Smith, D. I., Kaufmann, S. H., Roberts, L. R., Shridhar, V., 2003. Loss of HSulf-1 up-regulates heparin-binding growth factor signaling in cancer, *J Biol Chem*, 278, pp. 23107–17.

Lamanna, W. C., Frese, M. A., Balleininger, M., Dierks, T., 2008. Sulf loss influences N-, 2O-, and 6O-sulfation of multiple heparan sulfate proteoglycans and modulates FGF signaling, *J Biol Chem*, 283, pp. 27724–35.

Lao, D. H., Yusoff, P., Chandramouli, S., Philp, R. J., Fong, C. W., Jackson, R. A., Saw, T. Y., Yu, C. Y., Guy, G. R., 2007. Direct binding of PP2A to Sprouty2 and phosphorylation changes are a prerequisite for ERK inhibition downstream of fibroblast growth factor receptor stimulation, *J Biol Chem*, 282, pp. 9117–26.

Laufer, E., Nelson, C. E., Johnson, R. L., Morgan, B. A., Tabin, C., 1994. Sonic hedgehog and Fgf-4 act through a signaling cascade and feedback loop to integrate growth and patterning of the developing limb bud, *Cell*, 79, pp. 993–1003.

Lea, R., Papalopulu, N., Amaya, E., Dorey, K., 2009. Temporal and spatial expression of FGF ligands and receptors during *Xenopus* development, *Dev Dyn*, 238, pp. 1467–79.

Lecaudey, V., Cakan-Akdogan, G., Norton, W. H., Gilmour, D., 2008. Dynamic Fgf signaling couples morphogenesis and migration in the zebrafish lateral line primordium, *Development*, 135, pp. 2695–705.

Levenstein, M. E., Ludwig, T. E., Xu, R. H., Llanas, R. A., VanDenHeuvel-Kramer, K., Manning, D., Thomson, J. A., 2006. Basic fibroblast growth factor support of human embryonic stem cell self-renewal, *Stem Cells*, 24, pp. 568–74.

Lewandoski, M., Sun, X., Martin, G. R., 2000. Fgf8 signalling from the AER is essential for normal limb development, *Nat Genet*, 26, pp. 460–63.

Lewis, T., Groom, L. A., Sneddon, A. A., Smythe, C., Keyse, S. M., 1995. XCL100, an inducible nuclear MAP kinase phosphatase from *Xenopus laevis*: its role in MAP kinase inactivation in differentiated cells and its expression during early development, *J Cell Sci*, 108 (Pt 8), pp. 2885–96.

Li, C., Scott, D. A., Hatch, E., Tian, X., Mansour, S. L., 2007. Dusp6 (Mkp3) is a negative feedback regulator of FGF-stimulated ERK signaling during mouse development, *Development*, 134, pp. 167–76.

Lin, X., Buff, E. M., Perrimon, N., Michelson, A. M., 1999. Heparan sulfate proteoglycans are essential for FGF receptor signaling during *Drosophila* embryonic development, *Development*, 126, pp. 3715–23.

Linker, C., Stern, C. D., 2004. Neural induction requires BMP inhibition only as a late step, and involves signals other than FGF and Wnt antagonists, *Development*, 131, pp. 5671–81.

Lombardo, A., Isaacs, H. V., Slack, J. M., 1998. Expression and functions of FGF-3 in *Xenopus* development, *Int J Dev Biol*, 42, pp. 1101–07.

Lundin, L., Larsson, H., Kreuger, J., Kanda, S., Lindahl, U., Salmivirta, M., Claesson-Welsh, L., 2000. Selectively desulfated heparin inhibits fibroblast growth factor-induced mitogenicity and angiogenesis, *J Biol Chem*, 275, pp. 24653–60.

Mao, J., McGlinn, E., Huang, P., Tabin, C. J., McMahon, A. P., 2009. Fgf-dependent Etv4/5 activity is required for posterior restriction of Sonic Hedgehog and promoting outgrowth of the vertebrate limb, *Dev Cell*, 16, pp. 600–06.

Marcelle, C., Wolf, J., Bronnerfraser, M., 1995. The in-vivo expression of the FGF receptor frek messenger-RNA in avian myoblasts suggests a role in muscle growth and differentiation, *Dev Biol*, 172, pp. 100–14.

Marchal, L., Luxardi, G., Thome, V., Kodjabachian, L., 2009. BMP inhibition initiates neural induction via FGF signaling and Zic genes, *Proc Natl Acad Sci USA*, 106, pp. 17437–42.

Maretto, S., Muller, P. S., Aricescu, A. R., Cho, K. W., Bikoff, E. K., Robertson, E. J., 2008. Ventral closure, headfold fusion and definitive endoderm migration defects in mouse embryos lacking the fibronectin leucine-rich transmembrane protein FLRT3, *Dev Biol*, 318, pp. 184–93.

Mariani, F. V., Ahn, C. P., Martin, G. R., 2008. Genetic evidence that FGFs have an instructive role in limb proximal-distal patterning, *Nature*, 453, pp. 401–05.

Marics, I., Padilla, F., Guillemot, J. F., Scaal, M., Marcelle, C., 2002. FGFR4 signaling is a necessary step in limb muscle differentiation, *Development*, 129, pp. 4559–69.

Marques, S. R., Lee, Y., Poss, K. D., Yelon, D., 2008. Reiterative roles for FGF signaling in the establishment of size and proportion of the zebrafish heart, *Dev Biol*, 321, pp. 397–406.

Martinez, S., Wassef, M., Alvarado-Mallart, R. M., 1991. Induction of a mesencephalic phenotype in the 2-day-old chick prosencephalon is preceded by the early expression of the homeobox gene en, *Neuron*, 6, pp. 971–81.

Martynoga, B., Morrison, H., Price, D. J., Mason, J. O., 2005. Foxg1 is required for specification of ventral telencephalon and region-specific regulation of dorsal telencephalic precursor proliferation and apoptosis, *Dev Biol*, 283, pp. 113–27.

Mason, I., 2007. Initiation to end point: the multiple roles of fibroblast growth factors in neural development, *Nat Rev Neurosci*, 8, pp. 583–96.

Mason, J. M., Morrison, D. J., Bassit, B., Dimri, M., Band, H., Licht, J. D., Gross, I., 2004. Tyrosine phosphorylation of Sprouty proteins regulates their ability to inhibit growth factor signaling: a dual feedback loop, *Mol Biol Cell*, 15, pp. 2176–88.

Mason, J. M., Morrison, D. J., Basson, M. A., Licht, J. D., 2006. Sprouty proteins: multifaceted negative-feedback regulators of receptor tyrosine kinase signaling, *Trends Cell Biol*, 16, pp. 45–54.

Mathieu, J., Griffin, K., Herbomel, P., Dickmeis, T., Strahle, U., Kimelman, D., Rosa, F. M., Peyrieras, N., 2004. Nodal and Fgf pathways interact through a positive regulatory loop and synergize to maintain mesodermal cell populations, *Development*, 131, pp. 629–41.

Metzger, R. J., Klein, O. D., Martin, G. R., Krasnow, M. A., 2008. The branching programme of mouse lung development, *Nature*, 453, pp. 745–50.

Metzger, R. J., Krasnow, M. A., 1999. Genetic control of branching morphogenesis, *Science*, 284, pp. 1635–39.

Meyers, E. N., Martin, G. R., 1999. Differences in left–right axis pathways in mouse and chick: functions of FGF8 and SHH, *Science*, 285, pp. 403–06.

Millimaki, B. B., Sweet, E. M., Dhason, M. S., Riley, B. B., 2007. Zebrafish atoh1 genes: classic proneural activity in the inner ear and regulation by Fgf and Notch, *Development*, 134, pp. 295–305.

Min, H., Danilenko, D. M., Scully, S. A., Bolon, B., Ring, B. D., Tarpley, J. E., DeRose, M., Simonet, W. S., 1998. Fgf-10 is required for both limb and lung development and exhibits striking functional similarity to *Drosophila* branchless, *Genes Dev*, 12, pp. 3156–61.

Mohammadi, M., Honegger, A. M., Rotin, D., Fischer, R., Bellot, F., Li, W., Dionne, C. A., Jaye, M., Rubinstein, M., Schlessinger, J., 1991. A tyrosine-phosphorylated carboxy-terminal peptide of the fibroblast growth factor receptor (Flg) is a binding site for the SH2 domain of phospholipase C-gamma 1, *Mol Cell Biol*, 11, pp. 5068–78.

Moreno, T. A., Kintner, C., 2004. Regulation of segmental patterning by retinoic acid signaling during *Xenopus* somitogenesis, *Dev Cell*, 6, pp. 205–18.

Morimoto-Tomita, M., Uchimura, K., Werb, Z., Hemmerich, S., Rosen, S. D., 2002. Cloning and characterization of two extracellular heparin-degrading endosulfatases in mice and humans, *J Biol Chem*, 277, pp. 49175–85.

Murai, K., Vernon, A. E., Philpott, A., Jones, P., 2007. Hes6 is required for MyoD induction during gastrulation, *Dev Biol*, 312, pp. 61–76.

Nagano, T., Takehara, S., Takahashi, M., Aizawa, S., Yamamoto, A., 2006. Shisa2 promotes the maturation of somitic precursors and transition to the segmental fate in *Xenopus* embryos, *Development*, 133, pp. 4643–54.

Nechiporuk, A., Raible, D. W., 2008. FGF-dependent mechanosensory organ patterning in zebrafish, *Science*, 320, pp. 1774–77.

Nentwich, O., Dingwell, K. S., Nordheim, A., Smith, J. C., 2009. Downstream of FGF during mesoderm formation in *Xenopus*: the roles of Elk-1 and Egr-1, *Dev Biol*, 336, pp. 313–26.

Neugebauer, J. M., Amack, J. D., Peterson, A. G., Bisgrove, B. W., Yost, H. J., 2009. FGF signalling during embryo development regulates cilia length in diverse epithelia, *Nature*, 458, pp. 651–54.

Nicholson, K. M., Anderson, N. G., 2002. The protein kinase B/Akt signalling pathway in human malignancy, *Cell Signal*, 14, pp. 381–95.

Niswander, L., Jeffrey, Martin, G. R., Tickle, C., 1994. A positive feedback loop coordinates growth and patterning in the vertebrate limb, *BMC Evol Biol*, 2009; 9, p. 226.

Niswander, L., Tickle, C., Vogel, A., Booth, I., Martin, G. R., 1993. FGF-4 replaces the apical ectodermal ridge and directs outgrowth and patterning of the limb, *Cell*, 75, pp. 579–87.

Nonaka, S., Shiratori, H., Saijoh, Y., Hamada, H., 2002. Determination of left–right patterning of the mouse embryo by artificial nodal flow, *Nature*, 418, pp. 96–99.

Nonaka, S., Tanaka, Y., Okada, Y., Takeda, S., Harada, A., Kanai, Y., Kido, M., Hirokawa, N., 1998. Randomization of left–right asymmetry due to loss of nodal cilia generating leftward flow of extraembryonic fluid in mice lacking KIF3B motor protein, *Cell*, 95, pp. 829–37.

Northrop, J. L., Kimelman, D., 1994. Dorsal–ventral differences in Xcad-3 expression in response to FGF-mediated induction in *Xenopus*, *Dev Biol*, 161, pp. 490–503.

Nutt, S. L., Dingwell, K. S., Holt, C. E., Amaya, E., 2001. *Xenopus* Sprouty2 inhibits FGF-mediated gastrulation movements but does not affect mesoderm induction and patterning, *Genes Dev*, 15, pp. 1152–66.

Oginuma, M., Niwa, Y., Chapman, D. L., Saga, Y., 2008. Mesp2 and Tbx6 cooperatively create periodic patterns coupled with the clock machinery during mouse somitogenesis, *Development*, 135, pp. 2555–62.

Olivera-Martinez, I., Storey, K. G., 2007. Wnt signals provide a timing mechanism for the FGF-retinoid differentiation switch during vertebrate body axis extension, *Development*, 134, pp. 2125–35.

Ong, S. H., Guy, G. R., Hadari, Y. R., Laks, S., Gotoh, N., Schlessinger, J., Lax, I., 2000. FRS2 proteins recruit intracellular signaling pathways by binding to diverse targets on fibroblast growth factor and nerve growth factor receptors, *Mol Cell Biol*, 20, pp. 979–89.

Ornitz, D. M., 2000. FGFs, heparan sulfate and FGFRs: complex interactions essential for development, *Bioessays*, 22, pp. 108–12.

Orr-Urtreger, A., Bedford, M. T., Burakova, T., Arman, E., Zimmer, Y., Yayon, A., Givol, D., Lonai, P., 1993. Developmental localization of the splicing alternatives of fibroblast growth factor receptor-2 (FGFR2), *Dev Biol*, 158, pp. 475–86.

Ota, S., Tonou-Fujimori, N., Yamasu, K., 2009. The roles of the FGF signal in zebrafish embryos analyzed using constitutive activation and dominant-negative suppression of different FGF receptors, *Mech Dev*, 126, pp. 1–17.

Paek, H., Gutin, G., Hebert, J. M., 2009. FGF signaling is strictly required to maintain early telencephalic precursor cell survival, *Development*, 136, pp. 2457–65.

Panitz, F., Krain, B., Hollemann, T., Nordheim, A., Pieler, T., 1998. The Spemann organizer-expressed zinc finger gene Xegr-1 responds to the MAP kinase/Ets-SRF signal transduction pathway, *EMBO J*, 17, pp. 4414–25.

Park, E. J., Watanabe, Y., Smyth, G., Miyagawa-Tomita, S., Meyers, E., Klingensmith, J., Camenisch, T., Buckingham, M., Moon, A. M., 2008. An FGF autocrine loop initiated in second heart field mesoderm regulates morphogenesis at the arterial pole of the heart, *Development*, 135, pp. 3599–610.

Paterno, G. D., Ryan, P. J., Kao, K. R., Gillespie, L. L., 2000. The VT+ and VT– isoforms of the fibroblast growth factor receptor type 1 are differentially expressed in the presumptive mesoderm of *Xenopus* embryos and differ in their ability to mediate mesoderm formation, *Journal of Biological Chemistry*, 275, pp. 9581–86.

Pellegrini, L., 2001. Role of heparan sulfate in fibroblast growth factor signalling: a structural view, *Curr Opin Struct Biol*, 11, pp. 629–34.

Pera, E. M., Ikeda, A., Eivers, E., De Robertis, E. M., 2003. Integration of IGF, FGF, and anti-BMP signals via Smad1 phosphorylation in neural induction, *Genes Dev*, 17, pp. 3023–28.

Piccolo, S., Sasai, Y., Lu, B., De Robertis, E. M., 1996. Dorsoventral patterning in *Xenopus*: inhibition of ventral signals by direct binding of chordin to BMP-4, *Cell*, 86, pp. 589–98.

Pirvola, U., Ylikoski, J., Trokovic, R., Hebert, J. M., McConnell, S. K., Partanen, J., 2002. FGFR1 is required for the development of the auditory sensory epithelium, *Neuron*, 35, pp. 671–80.

Pownall, M., Gustaffson, M., Emerson, C., 2002. Myogenic regulatory factors and the specification of muscle progenitors in vertebrate embryos, *Annual Review of Cell and Dev Biol*, 18, pp. 747–83.

Pownall, M. E., Isaacs, H. V., Slack, J. M., 1998. Two phases of Hox gene regulation during early *Xenopus* development, *Current Biology*, 8, pp. 673–76.

Pownall, M. E., Tucker, A. S., Slack, J. M. W., Isaacs, H. V., 1996. eFGF, Xcad3 and Hox genes form a molecular pathway that establishes the anteroposterior axis in *Xenopus*, *Development*, 122, pp. 3881–92.

Pye, D. A., Gallagher, J. T., 1999. Monomer complexes of basic fibroblast growth factor and heparan sulfate oligosaccharides are the minimal functional unit for cell activation, *J Biol Chem*, 274, pp. 13456–61.

Pye, D. A., Vives, R. R., Hyde, P., Gallagher, J. T., 2000. Regulation of FGF-1 mitogenic activity by heparan sulfate oligosaccharides is dependent on specific structural features: differential requirements for the modulation of FGF-1 and FGF-2, *Glycobiology*, 10, pp. 1183–92.

Qiao, J., Bush, K. T., Steer, D. L., Stuart, R. O., Sakurai, H., Wachsman, W., Nigam, S. K., 2001. Multiple fibroblast growth factors support growth of the ureteric bud but have different effects on branching morphogenesis, *Mech Dev*, 109, pp. 123–35.

Randi, A. M., Sperone, A., Dryden, N. H., Birdsey, G. M., 2009. Regulation of angiogenesis by ETS transcription factors, *Biochem Soc Trans*, 37, pp. 1248–53.

Raya, A., Belmonte, J. C., 2006. Left–right asymmetry in the vertebrate embryo: from early information to higher-level integration, *Nat Rev Genet*, 7, pp. 283–93.

Regad, T., Roth, M., Bredenkamp, N., Illing, N., Papalopulu, N., 2007. The neural progenitor-specifying activity of FoxG1 is antagonistically regulated by CKI and FGF, *Nat Cell Biol*, 9, pp. 531–40.

Regan, J. C., Concha, M. L., Roussigne, M., Russell, C., Wilson, S. W., 2009. An Fgf8-dependent bistable cell migratory event establishes CNS asymmetry, *Neuron*, 61, pp. 27–34.

Reifers, F., Walsh, E. C., Leger, S., Stainier, D. Y., Brand, M., 2000. Induction and differentiation of the zebrafish heart requires fibroblast growth factor 8 (fgf8/acerebellar), *Development*, 127, pp. 225–35.

Ren, Y., Li, Z., Rong, Z., Cheng, L., Li, Y., Wang, Z., Chang, Z., 2007. Tyrosine 330 in hSef is critical for the localization and the inhibitory effect on FGF signaling, *Biochem Biophys Res Commun*, 354, pp. 741–46.

Ribisi, S., Jr., Mariani, F. V., Aamar, E., Lamb, T. M., Frank, D., Harland, R. M., 2000. Ras-mediated FGF signaling is required for the formation of posterior but not anterior neural tissue in *Xenopus laevis*, *Dev Biol*, 227, pp. 183–96.

Rossant, J., Ciruna, B., Partanen, J., 1997. FGF signaling in mouse gastrulation and anteroposterior patterning, *Cold Spring Harbor Symposia on Quantitative Biology*, 62, pp. 127–33.

Sai, X., Ladher, R. K., 2008. FGF signaling regulates cytoskeletal remodeling during epithelial morphogenesis, *Curr Biol*, 18, pp. 976–81.

Saint-Jeannet, J. P., Huang, S., Duprat, A. M., 1990. Modulation of neural commitment by changes in target cell contacts in Pleurodeles waltl, *Dev Biol*, 141, pp. 93–103.

Sasai, Y., Lu, B., Steinbeisser, H., De Robertis, E. M., 1995. Regulation of neural induction by the Chd and Bmp-4 antagonistic patterning signals in *Xenopus*, *Nature*, 376, pp. 333–36.

Sasaki, A., Taketomi, T., Kato, R., Saeki, K., Nonami, A., Sasaki, M., Kuriyama, M., Saito, N., Shibuya, M., Yoshimura, A., 2003. Mammalian Sprouty4 suppresses Ras-independent ERK activation by binding to Raf1, *Nat Cell Biol*, 5, pp. 427–32.

Sato, T., Joyner, A. L., Nakamura, H., 2004. How does Fgf signaling from the isthmic organizer induce midbrain and cerebellum development? *Dev Growth Differ*, 46, pp. 487–94.

Sato, T., Nakamura, H., 2004. The Fgf8 signal causes cerebellar differentiation by activating the Ras-ERK signaling pathway, *Development*, 131, pp. 4275–85.

Schlessinger, J., Plotnikov, A. N., Ibrahimi, O. A., Eliseenkova, A. V., Yeh, B. K., Yayon, A., Linhardt, R. J., Mohammadi, M., 2000. Crystal structure of a ternary FGF-FGFR-heparin complex reveals a dual role for heparin in FGFR binding and dimerization, *Mol Cell*, 6, pp. 743–50.

Schulte-Merker, S., Smith, J. C., 1995. Mesoderm formation in response to Brachyury requires FGF signalling, *Curr Biol*, 5, pp. 62–67.

Seed, J., Hauschka, S. D., 1988. Clonal analysis of vertebrate myogenesis. VIII. Fibroblasts growth factor (FGF)-dependent and FGF-independent muscle colony types during chick wing development, *Dev Biol*, 128, pp. 40–49.

Shi, W., Peyrot, S. M., Munro, E., Levine, M., 2009. FGF3 in the floor plate directs notochord convergent extension in the Ciona tadpole, *Development*, 136, pp. 23–28.

Shim, K., Minowada, G., Coling, D. E., Martin, G. R., 2005. Sprouty2, a mouse deafness gene, regulates cell fate decisions in the auditory sensory epithelium by antagonizing FGF signaling, *Dev Cell*, 8, pp. 553–64.

Shimamura, K., Rubenstein, J. L. R., 1997. Inductive interactions direct early regionalization of the mouse forebrain, *Development*, 124, pp. 2709–18.

Shimizu, T., Bae, Y. K., Hibi, M., 2006. Cdx-Hox code controls competence for responding to Fgfs and retinoic acid in zebrafish neural tissue, *Development*, 133, pp. 4709–19.

Sivak, J. M., Petersen, L. F., Amaya, E., 2005. FGF signal interpretation is directed by Sprouty and Spred proteins during mesoderm formation, *Dev Cell*, 8, pp. 689–701.

Slack, J. M. W., Darlington, B. G., Heath, J. K., Godsave, S. F., 1987. Mesoderm induction in early *Xenopus* embryos by heparin-binding growth factors, *Nature*, 326, pp. 197–200.

Sleeman, M., Fraser, J., McDonald, M., Yuan, S., White, D., Grandison, P., Kumble, K., Watson, J. D., Murison, J. G., 2001. Identification of a new fibroblast growth factor receptor, FGFR5, *Gene*, 271, pp. 171–82.

Smith, J. C., Price, B. M. J., Green, J. B. A., Weigel, D., Herrmann, B. G., 1991. Expression of a *Xenopus* homolog of Brachyury (T) is an immediate-early response to mesoderm induction, *Cell*, 67, pp. 79–87.

Smith, W. C., Harland, R. M., 1992. Expression cloning of noggin, a new dorsalizing factor localized to the Spemann organizer in *Xenopus* embryos, *Cell*, 70, pp. 829–40.

Srivastava, D., 2006. Making or breaking the heart: from lineage determination to morphogenesis, *Cell*, 126, pp. 1037–48.

Standley, H. J., Zorn, A. M., Gurdon, J. B., 2001. eFGF and its mode of action in the community effect during *Xenopus* myogenesis, *Development*, 128, pp. 1347–57.

Steinberg, F., Zhuang, L., Beyeler, M., Kalin, R. E., Mullis, P. E., Brandli, A. W., Trueb, B., 2010. The FGFRL1 receptor is shed from cell membranes, binds fibroblast growth factors (FGFs), and antagonizes FGF signaling in *Xenopus* embryos, *J Biol Chem*, 285, pp. 2193–202.

Streit, A., Berliner, A. J., Papanayotou, C., Sirulnik, A., Stern, C. D., 2000. Initiation of neural induction by FGF signalling before gastrulation, *Nature*, 406, pp. 74–78.

Streit, A., Lee, K. J., Woo, I., Roberts, C., Jessell, T. M., Stern, C. D., 1998. Chordin regulates primitive streak development and the stability of induced neural cells, but is not sufficient for neural induction in the chick embryo, *Development*, 125, pp. 507–19.

Sugaya, N., Habuchi, H., Nagai, N., Ashikari-Hada, S., Kimata, K., 2008. 6-O-sulfation of heparan sulfate differentially regulates various fibroblast growth factor-dependent signalings in culture, *J Biol Chem*, 283, pp. 10366–76.

Summerbell, D., Lewis, J. H., Wolpert, L., 1973. Positional information in chick limb morphogenesis, *Nature*, 244, pp. 492–96.

Suzuki-Hirano, A., Sato, T., Nakamura, H., 2005. Regulation of isthmic Fgf8 signal by sprouty2, *Development*, 132, pp. 257–65.

Tada, M., Concha, M. L., Heisenberg, C. P., 2002. Non-canonical Wnt signalling and regulation of gastrulation movements, *Semin Cell Dev Biol*, 13, pp. 251–60.

Takahashi, K., Tanabe, K., Ohnuki, M., Narita, M., Ichisaka, T., Tomoda, K., Yamanaka, S., 2007.

Induction of pluripotent stem cells from adult human fibroblasts by defined factors, *Cell*, 131, pp. 861–72.

Tanaka, Y., Okada, Y., Hirokawa, N., 2005. FGF-induced vesicular release of Sonic hedgehog and retinoic acid in leftward nodal flow is critical for left–right determination, *Nature*, 435, pp. 172–77.

Tao, Q., Yokota, C., Puck, H., Kofron, M., Birsoy, B., Yan, D., Asashima, M., Wylie, C. C., Lin, X., Heasman, J., 2005. Maternal wnt11 activates the canonical wnt signaling pathway required for axis formation in *Xenopus* embryos, *Cell*, 120, pp. 857–71.

Thisse, B., Thisse, C., 2005. Functions and regulations of fibroblast growth factor signaling during embryonic development, *Dev Biol*, 287, pp. 390–402.

Tickle, C., Wolpert, L., 2002. The progress zone—alive or dead? *Nat Cell Biol.* 4, pp. E216–17.

Towers, M., Tickle, C., 2009. Growing models of vertebrate limb development, *Development*, 136, pp. 179–90.

Trueb, B., Neuhauss, S. C., Baertschi, S., Rieckmann, T., Schild, C., Taeschler, S., 2005. Fish possess multiple copies of fgfrl1, the gene for a novel FGF receptor, *Biochim Biophys Acta*, 1727, pp. 65–74.

Tsang, M., Friesel, R., Kudoh, T., Dawid, I. B., 2002. Identification of Sef, a novel modulator of FGF signalling, *Nat Cell Biol*, 4, pp. 165–69.

Turnbull, J., Powell, A., Guimond, S., 2001. Heparan sulfate: decoding a dynamic multifunctional cell regulator, *Trends Cell Biol*, 11, pp. 75–82.

Turner, N., Grose, R., 2010. Fibroblast growth factor signalling: from development to cancer, *Nat Rev Cancer*, 10, pp. 116–29.

Ueda, Y., Hirai, S., Osada, S., Suzuki, A., Mizuno, K., Ohno, S., 1996. Protein kinase C activates the MEK-ERK pathway in a manner independent of Ras and dependent on Raf, *J Biol Chem*, 271, pp. 23512–19.

Ueno, H., Gunn, M., Dell, K., Tseng, A. J., Williams, L. T., 1992. A truncated form of fibroblast growth factor receptor 1 inhibits signal transduction by multiple types of fibroblast growth factor receptor, *Journal of Biological Chemistry*, 267, pp. 1470–76.

Umbhauer, M., Marshall, C. J., Mason, C. S., Old, R. W., Smith, J. C., 1995. Mesoderm induction in *Xenopus* caused by activation of MAP kinase, *Nature*, 376, pp. 58–62.

Urban, A. E., Zhou, X., Ungos, J. M., Raible, D. W., Altmann, C. R., Vize, P. D., 2006. FGF is essential for both condensation and mesenchymal-epithelial transition stages of pronephric kidney tubule development, *Dev Biol*, 297, pp. 103–17.

van Amerongen, R., Nusse, R., 2009. Towards an integrated view of Wnt signaling in development, *Development*, 136, pp. 3205–14.

Wahl, M. B., Deng, C., Lewandoski, M., Pourquie, O., 2007. FGF signaling acts upstream of the

NOTCH and WNT signaling pathways to control segmentation clock oscillations in mouse somitogenesis, *Development*, 134, pp. 4033–41.

Wang, S., Ai, X., Freeman, S. D., Pownall, M. E., Lu, Q., Kessler, D. S., Emerson, C. P., Jr., 2004. QSulf1, a heparan sulfate 6-O-endosulfatase, inhibits fibroblast growth factor signaling in mesoderm induction and angiogenesis, *Proc Natl Acad Sci USA*, 101, pp. 4833–38.

Wang, Y., Janicki, P., Koster, I., Berger, C. D., Wenzl, C., Grosshans, J., Steinbeisser, H., 2008. *Xenopus* Paraxial Protocadherin regulates morphogenesis by antagonizing Sprouty, *Genes Dev*, 22, pp. 878–83.

Wassarman, K. M., Lewandoski, M., Campbell, K., Joyner, A. L., Rubenstein, J. L., Martinez, S., Martin, G. R., 1997. Specification of the anterior hindbrain and establishment of a normal mid/hindbrain organizer is dependent on Gbx2 gene function, *Development*, 124, pp. 2923–34.

Weeks, D. L., Melton, D. A., 1987. A maternal mRNA localised to the vegetal hemisphere in Xenopus eggs codes for a growth factor related to TGF-ß, *Cell*, 51, pp. 861–67.

Weintraub, H., 1993. The myoD family and myogenesis redundancy, networks and thresholds, *Cell*, 75, 1241–44.

White, A. C., Xu, J., Yin, Y., Smith, C., Schmid, G., Ornitz, D. M., 2006. FGF9 and SHH signaling coordinate lung growth and development through regulation of distinct mesenchymal domains, *Development*, 133, pp. 1507–17.

Willert, K., Nusse, R., 1998. Beta-catenin: a key mediator of Wnt signaling, *Curr Opin Genet Dev*, 8, pp. 95–102.

Wills, A. E., Choi, V. M., Bennett, M. J., Khokha, M. K., Harland, R. M., 2009. BMP antagonists and FGF signaling contribute to different domains of the neural plate in *Xenopus*, *Dev Biol*, 337, pp. 335–50.

Wilson, S. I., Graziano, E., Harland, R., Jessell, T. M., Edlund, T., 2000. An early requirement for FGF signalling in the acquisition of neural cell fate in the chick embryo, *Curr Biol*, 10, pp. 421–29.

Winterbottom, E. F., Pownall, M. E., 2009. Complementary expression of HSPG 6-O-endosulfatases and 6-*O*-sulfotransferase in the hindbrain of *Xenopus laevis*, *Gene Expr Patterns*, 9, pp. 166–72.

Wright, T. J., Mansour, S. L., 2003. Fgf3 and Fgf10 are required for mouse otic placode induction, *Development*, 130, pp. 3379–90.

Yamamoto, A., Amacher, S. L., Kim, S. H., Geissert, D., Kimmel, C. B., De Robertis, E. M., 1998. Zebrafish paraxial protocadherin is a downstream target of spadetail involved in morphogenesis of gastrula mesoderm, *Development*, 125, pp. 3389–97.

Yamamoto, A., Nagano, T., Takehara, S., Hibi, M., Aizawa, S., 2005. Shisa promotes head formation through the inhibition of receptor protein maturation for the caudalizing factors, Wnt and FGF, *Cell*, 120, pp. 223–35.

Yang, X., Dormann, D., Munsterberg, A. E., Weijer, C. J., 2002. Cell movement patterns during gastrulation in the chick are controlled by positive and negative chemotaxis mediated by FGF4 and FGF8, *Dev Cell*, 3, pp. 425–37.

Yin, Y., White, A. C., Huh, S. H., Hilton, M. J., Kanazawa, H., Long, F., Ornitz, D. M., 2008. An FGF-WNT gene regulatory network controls lung mesenchyme development, *Dev Biol*, 319, pp. 426–36.

Yu, K., Ornitz, D. M., 2008. FGF signaling regulates mesenchymal differentiation and skeletal patterning along the limb bud proximodistal axis, *Development*, 135, pp. 483–91.

Zaffran, S., Kelly, R. G., Meilhac, S. M., Buckingham, M. E., Brown, N. A., 2004. Right ventricular myocardium derives from the anterior heart field, *Circ Res*, 95, pp. 261–68.

Zhang, X., Ibrahimi, O. A., Olsen, S. K., Umemori, H., Mohammadi, M., Ornitz, D. M., 2006. Receptor specificity of the fibroblast growth factor family. The complete mammalian FGF family, *J Biol Chem*, 281, pp. 15694–700.

Zhang, Z., Verheyden, J. M., Hassell, J. A., Sun, X., 2009. FGF-regulated Etv genes are essential for repressing Shh expression in mouse limb buds, *Dev Cell*, 16, pp. 607–13.

Zimmerman, L. B., De Jesus-Escobar, J. M., Harland, R. M., 1996. The Spemann organizer signal noggin binds and inactivates bone morphogenetic protein 4, *Cell*, 86, pp. 599–606.

About the Authors

Dr. M. E. Pownall and **Dr. H. V. Isaacs** have been running independent research groups in the Biology Department at the University of York in the UK since 1999. Dr. Pownall did her graduate work at the University of Virginia with Prof. Charles P. Emerson, Jr., where she cloned and characterized quail orthologues of newly identified myogenic regulatory factors. From there, she went on to do postdoctoral work with Prof. Jonathan Slack in Oxford where her interest in FGF signalling and training in amphibian embryology began. Dr. Isaacs also trained and worked in the Slack lab and was involved in the very early studies that involved purifying FGF protein from bovine brains. During their time together in the Slack lab, they identified an autoregulatory loop of FGF4 and Brachyury that maintains the mesoderm, and they began to elucidate a role for FGF in posteriorizing the early embryonic axis. Currently, both research groups are working to understand the regulation of FGF signalling and to characterize genes that are activated downstream of FGF during early amphibian development. The authors also share an interest in the well-being of their two children and large golden retriever.

About the Author